BIOCHEMISTRY RESEARCH TRENDS

ARACHIDONIC ACID

SOURCES, BIOSYNTHESIS AND HEALTH EFFECTS

BIOCHEMISTRY RESEARCH TRENDS

Additional books in this series can be found on Nova's website
under the Series tab.

Additional e-books in this series can be found on Nova's website
under the e-book tab.

BIOCHEMISTRY RESEARCH TRENDS

ARACHIDONIC ACID

SOURCES, BIOSYNTHESIS AND HEALTH EFFECTS

JASON M. O'KEEFE
EDITOR

nova publishers
New York

For permission to use material from this book please contact us:
Telephone 631-231-7269; Fax 631-231-8175
Web Site: http://www.novapublishers.com

Library of Congress Cataloging-in-Publication Data

ISBN: 978-1-63117-619-7

Published by Nova Science Publishers, Inc. † New York

CONTENTS

PREFACE

Arachidonic acid (ARA) is one the two main polyunsaturated fatty acids (PUFAs) in the brain, especially in neuronal cells, and can be a target in the fight against the major public health concern represented by Alzheimer's disease (AD) and related disorders in high- as well as in low-income countries where elders constitute a growing part of the population. This compilation includes topics such as the use of arachidonic acid to prevent Alzheimer's disease; the dietary effects and sources of ARA; the effect of ARA on key oncogenic pathways in prostate cancer; and the effects of adipokines on prostaglandin E2 production by rheumatoid synovial fibroblasts.

Chapter 1 - The effects of dietary arachidonic acid (ARA) on survival, growth, and hypersaline stress resistance of southern flounder larvae were determined from first feeding through metamorphosis. Larvae were fed rotifers and *Artemia* nauplii enriched with emulsions containing 10% docosahexaenoic acid (DHA) and five different levels of ARA (0, 1, 2, 4, and 6%). Larvae in a sixth treatment were fed unenriched rotifers cultured with *Nannochloropsis oculata* (0% DHA, 31% eicosapentaenoic acid, EPA and 3.94% ARA).

There were no significant differences ($P > 0.05$) among treatments in larval survival (41.9% ± 6.3), growth (notochord length = 6.8 mm ± 0.2, dry weight = 0.5 mg ± 0.06) or median survival time (41 min.) following hypersaline (60 g L^{-1}) challenge on d26 -d27ph. ARA and DHA levels in d27ph larvae reflected those of live prey, except for larvae fed unenriched prey (0% DHA), which contained 2.5% DHA, but with no diminution in performance. Results indicated that DHA and ARA prey enrichment was not required for southern flounder larvae fed rotifers cultured with *N. oculata,* suggesting sufficient basal levels of ARA in live prey and derivation of DHA from maternally-derived yolk reserves, or via biosynthesis from the precursor fatty acids α-linolenic acid (ALA), EPA, or docosapentaenoic acid (DPA).

Chapter 2 - Dietary habits in Western societies, in particular the amount and type of fat consumed, play an important role in human health condition. The present chapter reviews the biosynthesis, dietary sources and health implications of the arachidonic acid (20:4*n*-6, AA), a long-chain polyunsaturated fatty acid (PUFA) from the *n*-6 family found in mammalian cells, especially in membrane phospholipids. Systemic and tissue levels of AA may result from two origins, the endogenous synthesis from linoleic acid (18:2*n*-6, LA) elongation and the exogenous dietary source. LA elongation involves desaturation and chain elongation enzymes, which are also common to the *n*-3 PUFA α-linolenic acid (18:3*n*-3, ALA) elongation pathway. As dietary intake of ALA is low, the production of AA from LA is little restrained. Humans consume AA from animal-derived foods (e.g., eggs and meat) as it does not occur in plants. The major biological functions of AA are as cell signaling molecule and as precursor to an extensive group of compounds, called eicosanoids. Different eicosanoids, namely prostaglandins, thromboxanes, leukotrienes, and other bioactive metabolites, are produced by different types of cells via distinct enzymatic pathways. These AA mediators act like potent chemical messengers on immunity regulation. The overproduction of prothrombotic and pro-inflammatory eicosanoids, thromboxanes and leukotrienes derived from AA, has been associated with numerous diseases, such as cancer, heart disease, rheumatoid arthritis or asthma. In addition, AA and its end-products are involved in physiological processes in the central nervous system and have been implicated in neurological, neurodegenerative, and psychiatric disorders. During several decades, AA and its metabolites have been considered nothing but pro-inflammatory with recommendations to lower its intake from the diet. However, the discovery of anti-inflammatory mediators produced from AA is changing the long-held misunderstandings about this fatty acid. Unlike the pro-inflammatory prostaglandins and leukotrienes, AA-derived lipoxins display counter-regulatory actions by promoting the resolution of acute inflammatory responses. In conclusion, AA has been the focus of new investigations in the recent years and, rather than being detrimental, moderate AA intake seems to be a key dietary component for cardiovascular and cognitive health, with harmless effects in healthy subjects.

Chapter 3 - Arachidonic acid (ARA) is one of the two most abundant polyunsaturated fatty acids (PUFAs) in brain, especially in neuronal cells, and can be a target in the fight against the major public health concern represented by Alzheimer's disease (AD) and related disorders in high- as well as in low-income countries where elders constitute a growing part of the population. The numerous studies devoted to the AD pathophysiological mechanisms emphasized the role of the amyloid-β (Aβ) oligomers generated from the precursor protein APP. Aβ oligomer-induced disruption of hippocampal synapses is recognized as the first cellular alteration in AD, leading to reduced cognitive abilities. Furthermore, the occurrence of neuroinflammation and oxidative stress is strongly linked to the AD process. But the mechanisms responsible for AD initiation are still largely unknown, although their identification is critical to design efficient therapeutic or preventive strategies. ARA was first considered in AD as a mediator of neuroinflammation and oxidative stress through its conversion into pro-inflammatory eicosanoids by several enzymes such as 5-lipoxygenase and cyclooxygenase-2. Upregulation of these enzymes in AD brain could contribute to the pathophysiological process by amplifying neuroinflammation and by increasing Aβ oligomer production and neurotoxicity. Other studies showed that free ARA is involved in essential synaptic functions through its effects on several protein kinases or its interactions with SNARE components in the traffic of synaptic vesicles. Therefore, free intracellular ARA levels must be adjusted to physiological needs at adequate cellular locations. ARA is specifically released from membrane phospholipids by cytosolic phospholipase A_2 ($cPLA_2$-α). The authors and other groups demonstrated that $cPLA_2$-α is an upstream effector of the Aβ oligomer-induced apoptotic cascade, while neuroprotective effects are observed in $cPLA_2$-α KO models or in response to pharmacological inhibitors. Increase in intraneuronal free ARA levels could also result from the catabolism of endocannabinoids by monoacylglycerol lipase (MAGL). The modification of the activities of MAGL and other endocannabinoid-hydrolyzing enzymes in AD could reduce the putative positive effects of the endocannabinoid signaling. Interestingly, several acyl-CoA synthetases or lysophospholipid acyltransferases were recently characterized and could antagonize the effects of $cPLA_2$-α and MAGL on the free ARA levels in brain cells. Furthermore, ARA is believed to compete with incorporation of protective docosahexaenoic acid (DHA) into the membrane phospholipids. The unbalance between omega-3 and omega-6 PUFAs in Western diets potentially favors the increase in brain ARA pool compared to that of DHA. All these findings suggest that controlling the levels of intracellular free ARA could be of valuable interest in therapeutic or preventive strategies against AD, as discussed in this chapter.

Chapter 4 - Arachidonic acid is also subjected to reactions which leads to the formation of compounds known as leukotrienes. The first step in the formation of leukotrienes is cleavage of arachidonic acid from phospholipids membrane. The synthesis of leukotrienes begins with the action of several lipoxygenases, which catalyze the incorporation of molecular oxygen in arachidonic acid.

The arachidonic acid (AA) can also be converted by prostaglandin (PG) G/H synthase cytosolic, also known as cyclooxygenase (COX) in intermediates PGG2 and PGH2. The enzyme prostaglandin G/H synthase, has two sites catalyst: the site and the site cyclooxygenase peroxidase. The site cyclooxygenase converts the AA in PGG2, which in turn is reduced to the unstable intermediate, PGH2 by the peroxidase site, which is not inhibited by nonsteroidal anti-inflammatory drug (NSAIDS).

For NSAIDS treatment has been difficult to assign relative gastrointestinal "safety", since although they aimed COX-2 inhibition, the therapeutic doses of these drugs will also result in COX-1 reduced activity and severe hepatotoxicity. Therefore, there is a need to search for new therapeutic drugs used in the inhibition of COX-2 and other enzymes or inflammatory cytokines.

Some pharmacological studies showed that diethylcarbamazine (DEC) interferes with the arachidonic acid metabolism, acting as an anti-inflammatory drug. DEC is the drug most widely used in the treatment of lymphatic filariasis since 1947. Some studies performed on vertebrates show that this drug has several direct biochemical effects on different enzyme systems, including glycolysis, folate metabolism and activity of acetylcholinesterase.

DEC also has anti-inflammatory properties as a result of its interference with the arachidonic acid metabolism, which includes lipoxygenases (LOX) and cyclooxygenases (COX) enzymes.

For the first time, was demonstrated that DEC can decrease liver injury in experimental alcoholism and malnutrition model, and revealed a clinical potential of DEC for therapeutic anti-inflammatory and hepato-protective applications.

Chapter 5 - The dietary arachidonic acid (AA) is stored at the *sn-2* position in the glycerolphospholipid of cell membrane. AA is released *via* the cleavage by phospholipase A_2 in response to stimuli such as cytokines, growth factors and hormones. The freed AA is metabolised typically by cyclooxygenase and lipoxygenase enzymes to form eicosanoids. Of all the phospholipase A_2 that catalyse the hydrolysis of fatty acid at *sn-2* bond of glycerophospholipids, cytosolic phospholipase $A_2\alpha$ ($cPLA_2\alpha$) is preferentially targeting AA-containing phospholipids. Hence, cPLA$_2$α action represents the key step for AA release and eicosanoid synthesis. Stimulation of cells with agents that mobilise intracellular calcium and/or promote the phosphorylation of cPLA$_2$α leads to increased AA release and eicosanoid production.

AA and its metabolites have been shown to promote cancer cell proliferation and thus are under intensive investigation. Interestingly, the rate of eicosanoid production is also significantly higher in malignant than benign prostate tissues. AA increases the proliferation of prostate cancer cells *in vitro*. Mice on a diet with high AA content have more frequent and larger tumour recurrences following tumour removal. AA metabolites have been shown to affect cancer cell proliferation, invasion, angiogenesis and tumorigenesis. Blocking cPLA$_2$α inhibits proliferation of prostate cancer cells *in vitro* and *in vivo*. The mechanisms underlying AA stimulatory action on prostate cancer cell proliferation are likely due to an AA-mediated increase in activation of protein kinase B (AKT), extra-cellular regulated kinase (ERK) and androgen receptor (AR) signalling pathways.

Chapter 6 - Eicosanoids, including prostaglandins (PGs), leukotrienes (LTs), and lipoxins (LXs), regulate a wide variety of physiological responses and pathological processes. Among them, PGE$_2$ is one of the key molecules, especially in relation to inflammation and immunity. Rheumatoid arthritis (RA) is characterized by extensive inflammation and proliferation of the synovium in various joints of the body. Since proinflammatory cytokines, such as tumor necrosis factor-α (TNFα) and interleukin (IL)-1β, induce PGE$_2$ production by synovial fibroblasts, joint inflammation in RA is at least partly mediated by overproduction of PGE$_2$. The authors previously reported that PGE$_2$ overproduction led to a vicious cycle of inflammation because PGE$_2$ itself promotes the increased expression of microsomal prostaglandin E synthase-1 (mPGES-1), a downstream enzyme of cyclooxygenase-2 (COX-2).

Adipose tissue has long been considered to be merely a structural component of many organs and a site for energy storage. However, recent studies have demonstrated that the major cellular component of adipose tissue, the adipocyte, has the ability to synthesize and release various physiologically active molecules, including diponectin, leptin, and resistin, as well as well-known proinflammatory cytokines like TNFα and IL-6. These molecules are called adipokines. The authors recently found that adiponectin stimulates PGE$_2$ production by increasing the expression of COX-2 and mPGES-1 in rheumatoid synovial fibroblasts.

In this review, the authors summarize the significance of adipokines with regard to PGE_2 and rheumatoid inflammation. The authors also suggest that adiponectin might have a role as a proinflammatory cytokine in rheumatoid arthritis.

In: Arachidonic Acid
Editor: Jason M. O'Keefe

ISBN: 978-1-63117-619-7

Chapter 1

The Effects of Dietary Arachidonic Acid (20:4n-6) on Growth, Survival and Hypersaline Stress Resistance in Larval Southern Flounder *Paralichthys lethostigma* (Jordan and Gilbert 1884)

Wade O. Watanabe[1,*], Troy C. Rezek[1,†], Moti Harel[2,‡] and Pamela J. Seaton[3,§]

[1]Center for Marine Science, University of North Carolina Wilmington, Wilmington, NC, US
[2]Advanced Bionutrition Corp., Columbia, MD, US
[3]Department of Chemistry and Biochemistry, University of North Carolina Wilmington, Wilmington, NC, US

* Wade O. Watanabe: Center for Marine Science, University of North Carolina Wilmington, 601 South College Rd., Wilmington, NC. 28403-5927. E-mail: watanabew@uncw.edu.
† Corresponding author: Troy C. Rezek. Current Address: NOAA National Ocean Service, Center for Coastal Fisheries and Habitat Research, 101 Pivers Island Rd., Beaufort, NC 28516. E-mail: troy.rezek@noaa.gov.
‡ Moti Harel: Advanced Bionutrition Corp., 7155 Columbia Gateway Dr., Suite H, Columbia, MD 21406.
§ Pamela J. Seaton: Department of Chemistry and Biochemistry, University of North Carolina Wilmington, 601 South College Rd., Wilmington, NC 28403-5932.

Abstract

The effects of dietary arachidonic acid (ARA) on survival, growth, and hypersaline stress resistance of southern flounder larvae were determined from first feeding through metamorphosis. Larvae were fed rotifers and *Artemia* nauplii enriched with emulsions containing 10% docosahexaenoic acid (DHA) and five different levels of ARA (0, 1, 2, 4, and 6%). Larvae in a sixth treatment were fed unenriched rotifers cultured with *Nannochloropsis oculata* (0% DHA, 31% eicosapentaenoic acid, EPA and 3.94% ARA).

There were no significant differences ($P > 0.05$) among treatments in larval survival (41.9% ± 6.3), growth (notochord length = 6.8 mm ± 0.2, dry weight = 0.5 mg ± 0.06) or median survival time (41 min.) following hypersaline (60 g L^{-1}) challenge on d26 -d27ph. ARA and DHA levels in d27ph larvae reflected those of live prey, except for larvae fed unenriched prey (0% DHA), which contained 2.5% DHA, but with no diminution in performance. Results indicated that DHA and ARA prey enrichment was not required for southern flounder larvae fed rotifers cultured with *N. oculata,* suggesting sufficient basal levels of ARA in live prey and derivation of DHA from maternally-derived yolk reserves, or via biosynthesis from the precursor fatty acids α-linolenic acid (ALA), EPA, or docosapentaenoic acid (DPA).

Introduction

The nutritional aspects of essential fatty acids (EFA) in fish have been extensively studied with considerable attention focused on the n-3 polyunsaturated fatty acids (PUFA) requirements of marine fish, particularly eicosapentaenoic acid 20:5n-3 (EPA) and docosahexaenoic acid 22:6n-3 (DHA) (Watanabe, Izquierdo, Takeuchi, Satoh and Kitajima 1989; Takeuchi, Toyota, Satoh and Watanabe 1990; Watanabe 1993; Takeuchi, Masuda, Ishizaki, Watanabe, Kanematsu, Imaizumi and Tsukamoto 1996; Ibeas, Cejas, Fores, Badia, Gomes and Hernandez 1997). The essentiality of these fatty acids is based on their roles as necessary components of cell membranes and as precursors for prostaglandin hormones.

Most marine fish studied to date whose diets are rich in DHA and EPA are usually unable to chain elongate and further desaturate α-linolenic acid 18:3n-3 (ALA) to the more highly unsaturated EPA and DHA (Tacon, 1990; Bessonart, Izquierdo, Salhi, Hernandez-Cruz, Gonzalez and Fernandez-Palacios 1999). This is caused by a deficiency in one of two enzymes in the conversion pathway from ALA to EPA, i.e., the C18 to C20 elongase multienzyme complex or the Δ5 fatty acid desaturase (Bell and Sargent 2003). Since both EPA and DHA are required for good growth, they are considered as essential dietary requirements. However, in all of the larval marine fish species examined to date, DHA has been demonstrated to be superior to EPA in providing vitality to the larvae (Watanabe 1993). Essential fatty acid enrichment of live feed diets (rotifers and *Artemia* nauplii) is a routine aspect of larval marine fish aquaculture. Commercial preparations used to enrich live foods typically contain abundant amounts of DHA and lower levels of EPA (Koven, Barr, Lutzky, Ben-Atia, Weiss, Harel, Behrens and Tandler 2001).

Given that DHA is naturally found at high levels in neural tissue, it is thought to play a critical role in neural membrane structure and function, including visual acuity for optimum hunting success (Copeman, Parrish, Brown and Harel 2002). Elevated dietary EPA relative to DHA has been shown to have a negative impact on larval neural function leading to lower growth and survival (Copeman et al., 2002). In juvenile turbot *Scophthalmus maximus* (Linnaeus 1758), the most significant aspect of lipid metabolism in the developing brain was the accumulation of DHA, while no other fatty acid increased in percentage in total lipid (Mourente, Tocher and Sargent 1991). Futhermore, increased mortalities were shown in juvenile turbot when the DHA: EPA ratio in the diet was decreased from 13.8 to 2.2, without changing the percentage of total n-3 PUFA (Bell, Henderson, Pirie and Sargent 1985). Turbot larvae exhibited lower pigmentation success with lower DHA: EPA ratio in their diet (Rainuzzo, Reiten and Olsen 1997). In starved gilthead seabream *Sparus aurata* (Linnaeus 1758) larvae, DHA was highly conserved, indicating its particular importance during the larval growth stage (Koven, Kissil and Tandler 1989).

Another series of essential fatty acids, the n-6 series, are generally found in smaller proportion than the n-3 series in marine fish with a ratio of n-3: n-6 PUFA of 10 to 15:1 (Ackman, 1979) and have therefore received little attention in marine fish nutrition. Despite the relatively small amount of arachidonic acid (20:4n-6, ARA) represented in the tissue, it is also essential for marine carnivorous fish (Koven et al., 2001), because they are unable to chain elongate and further desaturate linoleic acid (18:2n-6, LA) to the highly unsaturated n-6 fatty acids such as ARA. Hence, enrichment of live foods with lipids rich in n-6 as well as n-3 fatty acids prior to feeding to marine fish larvae is necessary for optimal growth, survival and resistance to handling stress (Watanabe, Izquierdo, Takeuchi, Satoh, Kitajima 1983; Takeuchi et al., 1990; Koven et al., 2001; Carrier, Watanabe, Harel, Rezek, Seaton and Shafer

2011). The southern flounder is an important commercial fishery species in the US Atlantic coast and a promising candidate for aquaculture. While nutritional requirements of other larval flatfish species such as turbot and summer flounder *Paralichthys dentatus* (Linnaeus 1776) have been reported, relatively little published data is available on the nutritional needs of southern flounder larvae. The objectives of this study were to determine the effects of dietary arachidonic acid on survival, growth and stress resistance of southern flounder larvae raised from first feeding to the metamorphic stages.

METHODS

Experimental Animals

This study was conducted at the University of North Carolina Wilmington Center for Marine Science Aquaculture Facility, Wrightsville Beach, NC. Adult southern flounder were maintained in fiberglass recirculating (tank exchange = 1100% / day, makeup exchange = 10% / day) broodtanks (diameter = 2.5 m, depth = 1 m, volume = 4.9 m^3) under 34 g L^{-1} salinity and a controlled photothermal cycle (Watanabe, Woolridge and Daniels, 2006). Adult females at the post-vitellogenic stage of ovarian development were induced to spawn using luteinizing hormone releasing hormone – analogue (LHRH-a) (Berlinsky, King, Smith, Hamilton and Sullivan, 1996; Smith, McVey, Wallace 1999; Watanabe, Carroll and Daniels, 2001) at a dose of 50 μg kg^{-1} body weight. Floating (viable) eggs were separated from sinking (non-viable) eggs in a 2-L separatory funnel in 34 g L^{-1} seawater. Floating fertilized eggs were transferred to a 15-L separatory funnel before distribution to the rearing tanks.

Experimental System

The experimental system consisted of four temperature-controlled water baths (152 x 61 x 23 cm) each covered by a light hood (152 x 65 x 19 cm) and surrounded by a black polyethylene curtain. Each water bath contained eight cylindrical black plastic tanks (15-L working volume) in which the larvae were reared. Fresh water was continuously re-circulated through the baths and through a heater/chiller to maintain a constant temperature.

Illumination was provided by four 40-W fluorescent bulbs supplying full spectrum lighting. Light intensity was controlled by regulating the height of the hood and with shade cloth. Lights were equipped with automatic timers to control photoperiod.

Experimental Design

To study the effects of dietary ARA on southern flounder larval growth, survival and hypersaline stress resistance, first-feeding stage larvae were fed rotifers (*Brachionus rotundiformis)* and *Artemia franciscana* that were enriched in emulsions containing 10% DHA and five levels of ARA (0, 1, 2, 4 and 6%). In a sixth treatment, larvae were fed unenriched rotifers cultured with *Nannochloropsis oculata*, which contained no DHA and 3.94% ARA (Reed Mariculture, Campbell CA.) and newly hatched unenriched *Artemia* nauplii. Five replicate tanks were assigned to each treatment.

To begin the experiment, developing embryos were stocked using volumetric methods into tanks at 35 g L^{-1} salinity, a temperature of 17 °C and at a density of 47 embryos L^{-1}. Filtered (10-μm) UV-treated seawater was used for larval rearing experiments.

LIVE PREY ENRICHMENT

The DHA: ARA emulsions were formulated by combining DHAsco (DHA-rich single cell oil) and different ratios of ARAsco (ARA rich single cell oil) with olive oil and water with 5% lecithin, 1% ascorbic acid and 1% Tween-80 (with oil weight) to form a 1:1 oil/water mixture (Advanced Bionutrition, Columbia, MD, US). Emulsions were sealed and refrigerated for daily use. DHAsco and ARAsco are microbial-derived triglyceride oils that are rich in DHA and ARA, respectively. DHAsco was extracted from the heterotrophically grown algae *Crypthecodinium cohnii* and the ARAsco was extracted from the fungus *Mortirella alpina* (Martek Biosciences, Columbia, MD).

Prior to enrichment, rotifers were cultured in 150-L tanks under continuous illumination, 34 g L^{-1} and at temperatures of 22-25 °C. Rotifers were fed a diet of preserved microalgae *N. oculata* (Reed Mariculture, San Jose, California) at a rate of 0.75 g per million rotifers per day.

Artemia (Argent Laboratories, Redmond, WA, US) were hatched from decapsulated cysts 24 h before enrichment. During enrichment, rotifers and *Artemia* were held in 3-L beakers in aerated, UV-treated seawater at 34 g L^{-1}. Photoperiod and temperature were maintained at 24 L: 0 D and 25°C. Dissolved oxygen was maintained above 4 mg L^{-1}.

During enrichment, rotifer density was maintained at 500,000 L^{-1}. Rotifers were enriched for 8 h before feeding to the larvae. The emulsions were added at 0.1 g $500{,}000^{-1}$ rotifers at 0 h and 4 h. During the period of enrichment with DHA: ARA emulsions, the unenriched rotifers were fed preserved *N. oculata* at rate of 0.75 g per million rotifers.

During enrichment, *Artemia* density was maintained at 200,000 nauplii L^{-1}. *Artemia* were enriched for 16 h before feeding to the larvae. Emulsions were added at 0.3 g $250{,}000^{-1}$ nauplii at 0 h and 8 h. Samples of the enriched and unenriched rotifers and *Artemia* were collected on three separate days (N = 3) immediately after the enrichment period and stored at -20 °C until fatty acid analysis.

Feeding

On day 2 post hatch (d2ph), rotifers were added to the larval rearing tanks at a density of 5 mL^{-1}. Once feeding began, rotifer density was increased to 20 mL^{-1}. Rotifer density was maintained by quantifying the rotifers in each culture vessel using volumetric methods and adding the appropriate number of enriched rotifers to make up the difference on a daily basis. Background algae *N. oculata* was added daily to maintain a density of 300,000 cells mL^{-1}. On d22ph, newly hatched *Artemia* instar 1 nauplii were added at a density of 0.25 mL^{-1} in addition to enriched rotifers. By d24ph, rotifer feeding stopped and larvae were fed *Artemia* metanauplii enriched with the prescribed treatments at a density of 1 mL^{-1}.

Environmental Conditions

Temperature ($\bar{X} \pm$ SEM; 18.7 ± 0.03 °C), salinity (35 ± 0.05 g L^{-1}), dissolved oxygen (7.3 ± 0.01 mg L^{-1}), light intensity (410 ± 4.0 lux), and pH ($\bar{X}$ = 7.96, range = 7.8 - 8.1) were measured daily in each tank. Photoperiod was 18L: 6D (Moustakas, Watanabe and Copeland 2004).

Aeration was supplied at approximately 80-100 mL min^{-1} through diffusers (4 cm x 1.3 cm x 1.3 cm) placed on the bottom at the center of each tank (Mangino and Watanabe 2006). Once feeding began, 30-55% of water in each tank was exchanged daily, and the water surface was skimmed 3 times daily with a paper towel to remove oil films.

Growth and Survival

To monitor growth and survival, a minimum of 10 larvae were randomly sampled from each tank on d2, d5, d9, d12, d16, d19 and d26ph using volumetric methods. Larvae were anesthetized (0.3 g L^{-1} 2-phenoxyethanol in freshwater) then placed into a gridded 100 x 15-mm Petri dish. Living larvae were distinguished from dead ones by the presence or absence of a heartbeat as well as by opacity and appearance. Notochord length was measured with an ocular micrometer. Wet weights were recorded by weighing approximately 10 larvae on a microscope slide and then subtracting the weight of the slide using a Sartorius (Goettingen, Germany) electrobalance to the nearest 10 μg. To determine dry weights, larvae were dried at 60 °C until they reached a constant weight (72 h) and then re-weighed.

The larval density (number of larvae L^{-1}) was determined by dividing the number of larvae sampled by the volume of water collected. On d28ph, all remaining larvae were sampled for carcass fatty acid composition.

Fatty Acid Analysis

Sample preparation of live prey and larvae for fatty acid analysis was similar to Watanabe et al.,(1989). Samples were poured onto a 23-μm mesh screen which was blotted with a sponge from underneath to remove excess water. Samples (approx. 100-150 mg) were then transferred to a 1.2 mL cryogenic vial (Fisher Scientific US), filled with nitrogen, and then frozen (-20 °C) before lipid extraction and fatty acid analysis.

Total lipid and fatty acid analysis was conducted using a revised Folch, Lee and Stanley (1957) method using a chloroform/methanol mixture (1:1 v/v) to extract lipids from tissue. Total lipid wet weight was converted to total lipid dry weight (mg g^{-1}) based on the % dry weight values of larvae and prey of the same treatment groups. Lipid fatty acids were converted to their methyl esters (FAMEs) for GC analysis by the FAME reaction.

The first phase of the FAME reaction is a base catalyzed hydrolysis of the glycerol esters, which required the addition of 0.5 M NaOH/MeOH to the sample, and then heated (70-100 °C) for 30 minutes. The second phase is an esterification of the free fatty acids catalyzed by the addition of 1.5 mL of boron trifluoride-methanol (BF_3) and heated for an additional 30 minutes. FAMEs from all samples were identified using a Hewlett Packard 6890 gas chromatography flame-ionization-dectector (GC FID) (Hewlett Packard, US). Helium was used as the carrier gas. The column temperature profile was: 195 °C, hold for 8 min, increased to 270 at 15 °C min^{-1} and hold at 270 °C for 2 min (Rezek et al.,2010). FAME peaks were integrated using the HP Chemstation software package and individual FAMEs were identified by comparison of retention times to standards: GLC-84 (Nu Chek Prep US), eicosapentaenoic acid methyl ester, arachidonic acid methyl ester (Sigma-Aldrich US), steariodonic acid methyl ester (Cayman Chemical US).

Hypersaline Stress Resistance

On d13 and d27ph larvae were subjected to a hypersaline (60 g L^{-1}) challenge (Dhert, Lavens and Sorgeloos 1991). A minimum of 20 larvae were sampled from each rearing tank and transferred to a 100 x 15-mm Petri dish containing 50 mL of 60 g L^{-1} seawater at 20 °C. Survival was monitored at 5 min. intervals until complete mortality was observed. For each replicate tank, survival (%) was plotted over time and a second order polynomial regression was fitted to the data. Stress sensitivity index (ST50) was calculated as the time at which survival fell to 50%.

Analytical Methods

The effects of ARA on growth, survival, stress resistance and fatty acid composition were expressed as treatment means ± standard error of the mean (SEM). For percentage data, arcsine transformation was performed before analysis. To meet the assumptions of ANOVA, data were analyzed using Levene's test for homogeneity of variances. For data that met the assumptions of ANOVA, a Tukey-Kramer honesty significant difference test was used for multiple comparisons among means. If the data violated the assumptions of ANOVA a Kruskal-Wallis nonparametric pairwise comparison with Bonferoni adjustment factor was used for multiple comparisons among means. Analyses were performed using the JMP IN version 7 statistical software (SAS Institute Inc.).

Results

Fatty Acid Composition of Enriched Rotifers

Lipid and fatty acid composition of enriched rotifers are shown in Table 1. No significant differences ($P > 0.05$) were observed in total lipid (4.4 ± 0.68% or 255 ± 40.3 mg g^{-1} dry wt.). Saturated fatty acids (SFA) were significantly higher in unenriched rotifers (40.6 ± 1.3% TFA) compared to rotifers enriched with 10% DHA and 0, 1, 2, 4%, or 6% ARA (29.3 ± 1.1% TFA). Levels of monounsaturated fatty acids (MFA) in rotifers decreased significantly from 43.8 ± 1.5% TFA to 34.4 ± 0.4% TFA as levels of dietary ARA increased from 0 to 6% and were lowest in unenriched rotifers fed only *N. oculata* (25.3 ± 0.4% TFA).

Total n-3 PUFA's were significantly higher in unenriched rotifers (27.4 ± 1.0% TFA) compared to rotifers enriched with 10% DHA and 0, 1, 2, or 4% ARA (23.5 ± 0.65% TFA). Of the n-3 PUFA's, levels of ALA (5.8 ± 0.9% TFA), EPA (15.3 ± 0.5% TFA) and 22:5n-3 docosapentaenoic acid (DPA) (6.3 ± 0.4% TFA) were significantly higher in unenriched rotifers compared to all other treatments (2.38 ± 0.21%, 7.9 ± 1.0, 2.9 ± 0.2% TFA, respectively). Unenriched rotifers fed only *N. oculata* contained no detectable levels of DHA, while no significant differences in DHA levels were observed among rotifers enriched with 10% DHA and 0, 1, 2, 4, or 6% ARA (10.5 ± 1.1% TFA).

There was a trend toward higher total n-6 PUFA's as level of ARA enrichment increased, from 4.2 ± 0.3% at 0% ARA to 12.0 ± 0.6 % ARA at 6% ARA. Total n-6 PUFA's were significantly higher in rotifers enriched with 6% (12.0 ± 0.6% TFA) and 4% ARA (10.1 ± 0.7% TFA) compared to 2% ARA (7.4 ± 0.1% TFA), 1% ARA (5.4 ± 0.1% TFA) and 0% ARA (4.2 ± 0.3% TFA). Unenriched rotifers had an intermediate n-6 PUFA level of 8.6 ± 0.7%. Likewise, concentration of ARA increased from 1.5 ± 0.2% TFA at 0% ARA to 8.2 ± 0.5% TFA at 6% ARA, while unenriched rotifers contained an intermediate ARA level (3.5 ± 0.3% TFA).

Table 1. Total lipid and fatty acid composition (% of total fatty acid) (mean ± SEM, $N = 3$) of rotifers *Brachionus rotundiformis* enriched with five different oil emulsions consisting of 10% DHA and varying levels of ARA (0%, 1%, 2%, 4%, 6%) and an unenriched group

Rotifer Treatments (DHA:ARA)						
Fatty acids	10:0	10:1	10:2	10:4	10:6	Unenriched
14:0	3.70 ± 0.30	3.9 ± .2	4.5 ± 0.2	4.2 ± 0.1	4.7 ± 0.1	4.0 ± 0.7
16:0	15.3 ± 0.8[a]	14.6 ± 0.9[a]	17.6 ± 1.4[a]	17.2 ± 0.9[a]	17.3 ± 0.9[a]	25.1 ± 1.5[b]
16:1n-7	7.2 ± 0.9[a]	7.3 ± 0.9[a]	8.4 ± 1.2[a]	8.3 ± 1.1[a]	8.4 ± 1.2[a]	15.3 ± 1.6[b]
18:0	8.4 ± 0.3[b]	8.1 ± 0.2[ab]	6.3 ± 0.4[ab]	7.1 ± 0.5[ab]	5.7 ± 0.2[a]	8.2 ± 1.3[ab]
18:1n-9	35.9 ± 2.5[c]	36.1 ± 1.8[c]	30.3 ± 2.4[bc]	26.5 ± 1.5[b]	25.3 ± 1.4[b]	10.0 ± 1.2[a]
18:2n-6	2.7 ± 0.1[a]	2.8 ± 0.1[ab]	3.7 ± 0.1[abc]	4.2 ± 0.5[bc]	3.8 ± 0.1[abc]	5.1 ± 0.5[c]
18:3n-3	2.7 ± 0.3[a]	2.6 ± 0.2[a]	2.4 ± 0.2[a]	2.2 ± 0.2[a]	2.0 ± 0.1[a]	5.8 ± 0.9[b]
20:0	1.4 ± 0.7[a]	1.1 ± 0.3[a]	1.8 ± 0.2[ab]	1.8 ± 0.2[ab]	1.7 ± 0.1[ab]	3.3 ± 0.5[b]
20:1	0.7 ± 0.3	Undetected	Undetected	0.6 ± 0.3	0.7 ± 0.3	Undetected
20:4n-6	1.5 ± 0.2[a]	2.6 ± 0.1[b]	3.7 ± 0.1[b]	5.9 ± 0.5[c]	8.2 ± 0.5[d]	3.5 ± 0.3[b]
20:5n-3	7.4 ± 0.9[a]	7.2 ± 0.8[a]	8.2 ± 1.0[a]	8.4 ± 1.1[a]	8.1 ± 1.0[a]	15.3 ± 0.5[b]
22:5n-3	2.8 ± 0.3[a]	2.8 ± 0.1[a]	3.1 ± 0.3[a]	2.8 ± 0.3[a]	3.1 ± 0.1[a]	6.3 ± 0.4[b]
22:6n-3	10.4 ± 0.9	10.8 ± 0.8	10.0 ± 1.3	10.4 ± 1.6	11.1 ± 1.1	Undetected
∑SFA	28.8 ± 0.6[a]	27.7 ± 0.6[a]	30.2 ± 1.6[a]	30.4 ± 1.4[a]	29.3 ± 1.2[a]	40.6 ± 1.3[b]
∑MFA	43.8 ± 1.5[d]	43.4 ± 0.9[cd]	38.8 ± 1.3[bc]	35.3 ± 1.0[b]	34.4 ± 0.4[b]	25.3 ± 0.4[a]
∑n-3 PUFA	23.2 ± 0.9[a]	23.4 ± 0.3[a]	23.6 ± 0.4[a]	23.8 ± 1.0[a]	24.2 ± 0.4[ab]	27.4 ± 1.0[b]
∑n-6 PUFA	4.2 ± 0.3[a]	5.4 ± 0.1[ab]	7.4 ± 0.1[bc]	10.1 ± 0.7[de]	12.0 ± 0.6[e]	8.6 ± 0.7[cd]
n3/n6 PUFA	5.5 ± 0.3[d]	4.3 ± 0.1[c]	3.2 ± 0.1[b]	2.4 ± 0.2[ab]	2.0 ± 0.1[a]	3.2 ± 0.4[bc]
DHA/EPA	1.5 ± 0.3	1.6 ± 0.3	1.3 ± 0.4	1.4 ± 0.4	1.4 ± 0.3	na
DHA/ARA	7.1 ± 1.2[b]	4.2 ± 0.4[ab]	2.7 ± 0.3[a]	1.8 ± 0.1[a]	1.4 ± 0.1[a]	na
EPA/ARA	4.8 ± 0.3[c]	2.8 ± 0.4[b]	2.2 ± 0.3[ab]	1.5 ± 0.3[ab]	1.0 ± 0.2[a]	4.4 ± 0.4[c]
Total lipid %	5.3 ± 0.3	5.6 ± 1.2	4.0 ± 0.6	4.9 ± 0.4	3.9 ± 1.1	2.5 ± 0.5
Total lipid mg g^{-1}	311 ± 18.3	327 ± 67.8	233 ± 34.4	283 ± 23.9	228 ± 65.1	147 ± 32.1

Column titles refer to the percentage of DHA: ARA in each enrichment emulsion.
Fatty acids in trace amounts (< 0.5%) are not reported.
Treatment means are compared using ANOVA ($P < 0.05$), and a Tukey-Kramer HSD was used for multiple comparisons among means.
Means not sharing a common letter are significantly ($P < 0.05$) different.
Total lipid is reported in mg g^{-1} dry weight.
ANOVA, analysis of variance; ARA, arachidonic acid; DHA docosahexaenoic acid; PUFA, polyunsaturated fatty acids.

The ratio of n3/n6 PUFA was significantly higher in the rotifers enriched 10% DHA and 0% ARA (5.5 ± 0.3) than in all other treatments (3.02 ± 0.18).

No significant differences in the ratio of DHA/EPA (1.4 ± 0.3) were observed. The ratio of DHA/ARA was significantly higher in rotifers enriched with 10% DHA and 0% ARA (7.1 ± 1.2) than in rotifers enriched with 2, 4 and 6% ARA (2.0 ± 0.2).

The ratio of EPA/ARA was significantly higher in rotifers enriched with 10% DHA and 0% ARA (4.8 ± 0.3) and in unenriched (4.4 ± 0.4) rotifers than in those enriched with 1, 2, and 4% ARA (2.17 ± 0.3), while rotifers enriched with 6% ARA (1.0 ± 0.2) showed the lowest level (Table 1).

Fatty Acid Composition of Enriched *Artemia*

Lipid and fatty acid composition of enriched *Artemia* are shown in Table 2. No significant differences were observed in total lipid (3.6 ± 0.37% or 264 ± 27.1 mg g^{-1} dry wt.). Total SFA was significantly higher in the 4% ARA (19.8 ± 0.1% TFA) treatment compared to the 0, 1, 2, and 6% ARA (17.8 ± 0.4 TFA) and the unenriched (18.0 ± 0.1) treatments. A trend toward decreasing MFA with increasing dietary ARA was observed from 49.6 ± 2.0 TFA at 0% ARA to 41.1 ± 0.9 TFA at 6% ARA. Unenriched *Artemia* had an intermediate MFA of 31.8 ± 0.6%.

Total n-3 PUFA levels in enriched *Artemia* were significantly higher in the 4% ARA (36.3 ± 0.4% TFA) and in the unenriched (39.8 ± 0.1% TFA) treatments compared to the 0, 1, 2 and 6% ARA treatments (28.4 ± 0.9% TFA) (Table 2).

Of the n-3 PUFA's, ALA was significantly higher in unenriched *Artemia* (34.3 ± 0.4% TFA) compared to the enriched *Artemia* (20.5 ± 0.9 TFA). A small amount (1.2 ± 0.6% TFA) of 18:4n-3 stearidonic acid (STD) was detected in all treatments, except in *Artemia* fed 0% and 4% ARA. Levels of EPA were significantly higher in the unenriched *Artemia* (5.5 ± 0.3% TFA) than in *Artemia* enriched with 0, 1 and 4% ARA (3.4 ± 0.3% TFA), but not with 2 and 6% ARA (4.1 ± 0.2 TFA).

DHA levels in *Artemia* (range = 3.8 – 8.7% TFA) were considerably lower than provided in their emulsions diets (10%), with significant variation among treatments. DHA levels were significantly higher in *Artemia* enriched with 0, 1 and 4% ARA (7.1 ± 0.6 % TFA) than in those enriched with 2% (3.8 ± 0.7% TFA) and 6% ARA and (4.3 ± 0.4% TFA). DHA was not detected in unenriched *Artemia*.

Table 2. Total lipid and fatty acid composition (% of total fatty acid) (mean ± SEM, *N* = 3) of *Artemia* enriched with five different oil emulsions consisting of 10% DHA and varying levels of ARA (0%, 1%, 2%, 4%, 6%) and an unenriched group

	Artemia Treatments (DHA:ARA)					
Fatty acids	10:0	10:1	10:2	10:4	10:6	Unenriched
14:0	2.1 ± 0.3^{ab}	2.1 ± 0.3^{ab}	1.6 ± 0.1^{ab}	2.4 ± 0.5^{b}	1.6 ± 0.1^{ab}	1.1 ± 0.03^{a}
16:0	10.8 ± 0.3	10.8 ± 0.3	11.1 ± 0.4	10.6 ± 0.2	11.0 ± 0.1	11.7 ± 0.1
16:1n-7	3.8 ± 0.3^{ab}	4.0 ± 0.2^{ab}	4.6 ± 0.4^{ab}	3.7 ± 0.1^{a}	4.9 ± 0.2^{b}	6.4 ± 0.1^{c}
18:0	3.8 ± 1.1	5.4 ± 0.1	5.2 ± 0.1	6.1 ± 0.4	5.1 ± 0.1	5.2 ± 0.1
18:1n-9	45.8 ± 1.7^{c}	42.0 ± 0.7^{bc}	38.0 ± 2.2^{b}	29.9 ± 0.6^{a}	36.1 ± 1.1^{b}	25.4 ± 0.7^{a}
18:2n-6	4.7 ± 0.1^{a}	4.8 ± 0.2^{a}	5.3 ± 0.02^{a}	4.9 ± 0.1^{a}	5.5 ± 0.1^{ab}	6.5 ± 0.5^{b}
18:3n-3	17.4 ± 1.2^{a}	17.6 ± 0.2^{ab}	21.8 ± 1.5^{bc}	24.7 ± 0.3^{c}	20.9 ± 1.1^{abc}	34.3 ± 0.4^{d}
18:4n-3	Undetected	0.88± 0.4	1.2 ± 0.6	Undetected	1.1 ± 0.6	1.5 ± 0.9
20:4n-6	1.1 ± 0.1^{a}	1.9 ± 0.2^{b}	2.5 ± 0.3^{b}	5.2 ± 0.2^{c}	5.2 ± 0.6^{c}	2.1 ± 0.2^{b}
20:5n-3	3.5 ± 0.3^{a}	3.8 ± 0.4^{a}	4.2 ± 0.3^{ab}	3.0 ± 0.3^{a}	4.1 ± 0.1^{ab}	5.5 ± 0.3^{b}
22:6n-3	6.3 ± 0.3^{cd}	6.4 ± 0.3^{cd}	3.8 ± 0.7^{b}	8.7 ± 1.1^{d}	4.3 ± 0.4^{bc}	Undetected
∑SFA	16.9 ± 0.4^{a}	18.3 ± 0.5^{ab}	18.2 ± 0.5^{a}	19.8 ± 0.1^{b}	17.8 ± 0.3^{a}	18.0 ± 0.1^{a}
∑MFA	49.6 ± 2.0^{c}	46.1 ± 0.5^{bc}	42.5 ± 1.8^{b}	33.6 ± 0.5^{a}	41.1 ± 0.9^{b}	31.8 ± 0.6^{a}
∑n-3 PUFA	27.3 ± 1.7^{a}	27.8 ± 0.3^{a}	30.0 ± 0.7^{a}	36.3 ± 0.4^{b}	29.2 ± 0.6^{a}	39.8 ± 0.1^{b}
∑n-6 PUFA	5.6 ± 0.2^{a}	6.7 ± 0.3^{ab}	7.7 ± 0.3^{b}	10.0 ± 0.1^{cd}	10.7 ± 0.7^{d}	8.6 ± 0.7^{bc}
n3/n6 PUFA	4.8 ± 0.1^{c}	4.2 ± 0.1^{bc}	3.9 ± 0.2^{bc}	3.6 ± 0.1^{ab}	2.8 ± 0.2^{a}	4.7 ± 0.4^{c}
DHA/EPA	1.8 ± 0.1^{ab}	1.7 ± 0.3^{ab}	0.9 ± 0.1^{a}	3.1 ± 0.7^{b}	1.0 ± 0.1^{a}	na
DHA/ARA	6.2 ± 0.1^{c}	3.4 ± 0.4^{b}	1.5 ± 0.1^{a}	1.7 ± 0.2^{a}	0.8 ± 0.1^{a}	na
EPA/ARA	3.4 ± 0.1^{d}	1.2 ± 0.1^{b}	1.7 ± 0.1^{b}	0.6 ± 0.1^{a}	0.8 ± 0.1^{ab}	2.6 ± 0.1^{c}
Total lipid %	3.5 ± 0.2	2.8 ± 0.1	3.2 ± 0.4	4.1 ± 0.4	3.4 ± 0.6	4.3 ± 0.5
Total lipid mg g^{-1}	262 ± 13.3	211 ± 10.2	239 ± 31.8	302 ± 27.1	250 ± 43.0	321 ± 37.4

Column titles refer to the percentage of DHA:ARA in each enrichment emulsion. Fatty acids in trace amounts (< 0.5%) are not reported.

Treatment means are compared using ANOVA ($P < 0.05$), and a Tukey-Kramer HSD was used for multiple comparisons among means.

Means not sharing a common letter are significantly ($P < 0.05$) different.

Total lipid is reported in mg g^{-1} dry weight.

ANOVA, analysis of variance; ARA, arachidonic acid; DHA docosahexaenoic acid; PUFA, polyunsaturated fatty acids.

The n-6 PUFA's in *Artemia* increased with ARA levels in the emulsions and were significantly higher at the 4% (10.0 ± 0.1% TFA) and 6% (10.7 ± 0.7% TFA) ARA levels than in the 1% (6.6 ± 0.3% TFA) and 2% ARA (7.7 ± 0.3% TFA) levels and were lowest at 0% ARA (5.6 ± 0.2% TFA).

Unenriched *Artemia* had an intermediate level of n-6 PUFA's (8.6 ± 0.7 TFA). Likewise, concentration of ARA in enriched *Artemia* increased significantly with increasing level of ARA enrichment from 1.1 ± 0.1% TFA at 0% ARA to 5.2 ± 0.6% TFA at 6% ARA. Unenriched *Artemia* contained an intermediate ARA concentration (2.1 ± 0.2% TFA).

The ratio of n3/n6 PUFA significantly declined with increasing level of ARA enrichment from 4.8 ± 0.1 at 0% ARA to 2.8 ± 0.2 at 6% ARA. The n3/n6 PUFA ratio was significantly higher in *Artemia* enriched with 10% DHA and from 0% (4.8 ± 0.1) to 2% ARA (3.9 ± 0.2) than in those enriched with 6% ARA (2.8 ± 0.2). The n3/n6 ratio in unenriched *Artemia* (4.7 ± 0.4) was not significantly different from those enriched with 0% ARA. The DHA/EPA ratio was significantly higher in *Artemia* enriched with 10% DHA and 4% ARA (3.1 ± 0.7) than in *Artemia* enriched with 2% (0.9 ± 0.1) or 6% ARA (1.0 + 0.1), but was not different from *Artemia* enriched with 0% or 1% ARA (1.8 ± 0.2).

The ratio of DHA/ARA declined significantly as level of ARA enrichment increased, from 6.2 ± 0.1 in *Artemia* enriched with 10% DHA and 0% ARA to 0.8 ± 0.1 in those enriched with 6% ARA. The EPA/ARA ratio was significantly higher in *Artemia* enriched with 10% DHA and 0% ARA (3.4 ± 0.1) than in *Artemia* enriched with 1%, 2% and 6% ARA (1.2 ± 0.1), while lowest values were observed 4% ARA (0.6 ± 0.1). Unenriched *Artemia* had a relatively high EPA/ARA ratio of 2.6 ± 0.1, significantly lower than the 0% ARA treatment.

Fatty Acid Composition of D27ph Larvae

Whole body lipid and fatty acid composition of southern flounder larvae d27ph are shown in Table 3. No significant differences were observed in total lipid (3.5 ± 0.5% or 272 ± 42.4 mg g^{-1} dry wt.). Concentration of SFA was significantly higher in larvae fed the unenriched prey (30.9 ± 0.2% TFA) than in the other treatments (26.8 ± 0.5% TFA). Total MFA was not significantly different among larvae fed prey enriched with 0%, 1% or 2% ARA (29.2 ± 1.3% TFA). MFA was significantly higher in larvae fed prey enriched with 1% ARA (32.6 ± 0.6% TFA) than in the 4% and 6% ARA, or unenriched prey treatments (23.9 ± 1.1% TFA). Total n-3 PUFA levels in d27ph larval tissue were significantly lower in larvae fed prey enriched with 1% ARA (29.1 ± 1.0% TFA) than in all other treatments (33.7 ± 0.6% TFA).

Table 3. Total lipid and fatty acid composition (% of total fatty acid) (mean ± SEM, N = 5) of d27ph larvae fed enriched rotifers *Brachionus rotundiformis* and *Artemia*, using 5 different oil emulsions consisting of 10% DHA and varying levels of ARA (0%, 1%, 2%, 4%, 6%) and an unenriched group

	Enriched diets received by larvae (DHA:ARA)					
Fatty acids	10:0	10:1	10:2	10:4	10:6	Unenriched
14:0	1.0 ± 0.5	1.3 ± 0.1	1.3 ± 0.0	1.4± 0.1	1.0 ± 0.4	1.0 ± 0.5
16:0	16.4 ± 0.5	16.2 ± 0.1	16.5 ± 0.6	17.7 ± 0.3	17.2 ± 0.7	19.9 ± 1.8
16:1n-7	3.9 ± 0.3^{ab}	4.3 ± 0.1^{ab}	4.2 ± 0.2^{ab}	3.9 ± 0.0^{a}	2.6 ± 1.3^{b}	4.5 ± 0.1^{c}
18:0	5.7 ± 0.2	6.7 ± 0.7	5.7 ± 0.1	5.1 ± 0.1	6.0 ± 0.5	6.7 ± 0.7
18:1n-9	23.2 ± 1.1^{b}	25.8 ± 0.4^{ab}	21.9 ± 0.9^{ab}	20.3 ± 0.3^{ab}	22.2 ± 1.7^{ab}	18.1 ± 1.7^{a}
18:2n-6	8.2 ± 0.3^{ab}	7.7 ± 0.2^{a}	7.7 ± 0.3^{ab}	8.7 ± 0.1^{ab}	9.1 ± 0.4^{ab}	10.2 ± 1.1^{b}
18:3n-3	16.0 ± 1.8	13.3 ± 2.1	15.1 ±1.5	13.4 ± 0.2	13.6 ± 2.9	17.0 ± 1.6
20:0	3.2 ± 0.2	2.9 ± 0.2	3.1± 0.2	3.9 ± 0.5	2.1 ± 1.1	3.2 ± 1.8
20:4n-6	3.1 ± 0.3^{a}	3.3 ± 0.1^{b}	4.2 ± 0.4^{b}	5.7 ± 0.1^{c}	6.8 ± 0.8^{c}	3.9± 0.3^{b}
20:5n-3	6.0 ± 0.1^{a}	5.0 ± 0.0^{a}	6.1 ± 0.1^{ab}	5.0 ± 0.1^{a}	5.5 ± 0.2^{ab}	8.2 ± 0.4^{b}
22:5n-3	3.1 ± 0.4	2.7 ± 0.2	3.7 ± 0.4	3.2 ± 0.1	4.0 ± 0.6	4.6 ± 2.3
22:6n-3	9.3 ± 2.3^{cd}	8.1 ± 1.0^{cd}	9.4 ± 1.7^{b}	11.6 ± 0.1^{d}	10.7 ± 2.0^{bc}	2.6 ± 1.3^{a}
∑SFA	26.3 ± 0.2^{a}	27.2 ± 0.5^{a}	26.5 ± 0.3^{a}	28.1± 0.8^{a}	25.7± 0.7^{a}	30.9 ± 0.2^{b}
∑MFA	27.8± 1.6^{ab}	32.6 ± 0.6^{b}	27.1 ± 1.7^{ab}	24.3± 0.3^{a}	24.8 ± 1.2^{a}	22.6 ± 1.8^{a}
∑n-3 PUFA	34.5 ± 1.1^{b}	29.1 ± 1.0^{a}	34.3 ± 0.7^{b}	33.2 ± 0.5^{b}	33.9 ± 0.6^{b}	32.5 ± 1.0^{ab}
∑n-6 PUFA	11.4 ± 0.6^{a}	11.0 ± 0.1^{ab}	12.0 ± 0.7^{b}	14.4 ± 0.1^{cd}	16.1± 1.2^{d}	14.1 ± 1.4^{bc}
n3/n6	3.0 ± 0.1^{c}	2.7 ± 0.1^{abc}	2.9 ± 0.1^{bc}	2.3 ± 0.0^{ab}	2.2 ± 0.2^{a}	2.4 ± 0.3^{abc}
DHA/EPA	1.5 ± 0.4^{ab}	1.6 ± 0.2^{b}	1.5 ± 0.3^{ab}	2.3 ± 0.0^{b}	1.9 ± 0.4^{b}	0.3 ± 0.2^{a}
DHA/ARA	2.9 ± 0.4^{c}	2.5 ± 0.2^{bc}	2.2 ± 0.2^{bc}	2.0 ± 0.0^{bc}	1.5 ± 0.1^{ab}	0.7 ± 0.4^{a}
EPA/ARA	1.9 ± 0.2^{cd}	1.5 ± 0.1^{bc}	1.4 ± 0.1^{b}	0.9 ± 0.0^{a}	0.8 ± 0.1^{a}	2.1 ± 0.1^{d}
Total lipid %	3.5 ± 0.2	3.7 ± 0.1	3.0 ± 0.6	3.5± 0.7	3.5 ± 1.0	3.8 ± 0.5
Total lipid mg g^{-1}	207 ±12.8	296 ± 12.2	239 ± 43.9	286± 58.4	296 ± 83.2	309 ± 44.0

Column titles refer to the percentage of DHA: ARA in each enrichment emulsion. Fatty acids in trace amounts (< 0.5%) are not reported.

Treatment means are compared using ANOVA ($P < 0.05$), and a Tukey-Kramer HSD was used for multiple comparisons among means.

Means not sharing a common letter are significantly ($P< 0.05$) different.

Total lipid is reported in mg g^{-1} dry weight.

ANOVA, analysis of variance; ARA, arachidonic acid; DHA docosahexaenoic acid; PUFA, polyunsaturated fatty acids.

ALA ranged from 13.3 ± 2.1 to 17.0 ± 1.7% TFA among treatments, with no significant differences.

EPA levels were significantly higher in larvae fed unenriched prey (8.2 ± 0.4% TFA) than in the 0, 1, and 4% ARA treatments (5.3 ± 0.1% TFA), but not the 2 and 6% ARA (5.8 ± 0.2% TFA) treatments.

DHA concentration in larvae fed prey enriched with 10% DHA ranged from 8.1 ± 1.0% at 1% ARA to 11.6 ± 0.1% TFA at 4% ARA, significantly higher than in larvae fed unenriched prey (2.6 ± 1.3% TFA). DPA ranged from 2.7 ± 0.2 to 4.6 ± 2.3% TFA among treatments, with no significant differences.

The total n-6 PUFA concentration was significantly higher in larvae fed 4% and 6% ARA (15.3 ± 0.7% TFA) than in larvae fed 0%, 1% or 2% ARA (11.5 ± 0.5% TFA), or unenriched (14.1 ± 1.4% TFA) prey.

Concentration of ARA in the larval tissue increased with increasing dietary ARA and was significantly higher in larvae fed 4% and 6% ARA (6.3 ± 0.5% TFA) than in those fed 1% and 2% ARA or unenriched prey (3.8 ± 0.3% TFA) prey and lowest at 0% ARA (3.1 ± 0.3% TFA).

Growth

For the duration of the experiment (d2ph through d26ph), no significant treatment effects were found on larval notochord length, wet weight or dry weight (Table 4). By d26ph, mean notochord length, wet weight and dry weight among treatments were 6.77 ± 0.17 mm, 3.87 ± 0.21 mg and 0.51 ± 0.06 mg, respectively.

Survival and Hypersaline Stress Evaluation

There were no significant differences in survival among treatment groups from d2ph (71.6 ± 2.7%) to d26ph (41.9 ± 2.1%) (Table 5). On d26ph, mean survival in all treatment groups receiving rotifers and *Artemia* enriched with 10% DHA was 41.3 ± 2.4%, while survival in larvae fed unenriched prey was 44.9 ± 5.9%.

During the hypersaline challenge, median survival time (ST50) among larvae in all treatments averaged 12.6 ± 3.0 min on d13ph (Table 6), but increased to 40.2 ± 7.0 min on d27ph, with no significant differences.

Table 4. Notochord length (NL, mm), wet weight (WW, mg) and dry weight (DW, mg) (mean ± SEM, *N* = 5) of larvae fed rotifers *Brachionus rotundiformis* and *Artemia*, enriched with five different oil emulsions consisting of 10% DHA and varying levels of ARA (0%, 1%, 2%, 4%, 6%) and an unenriched group. No significant treatment differences were observed for any parameter on any sampling date

	Age (dph)	10:0	10:1	10:2	10:4	10:6	*N. oculata*
NL	2	3.42 ± 0.03	3.39 ± 0.04	3.37 ± 0.01	3.29 ± 0.02	3.36 ± 0.04	3.33 ± 0.01
WW	2	0.24 ± 0.02	0.24 ± 0.01	0.25 ± 0.01	0.25 ± 0.01	0.26 ± 0.01	0.26 ± 0.01
DW	2	0.02 ± 0.00	0.02 ± 0.00	0.02 ± 0.00	0.03 ± 0.00	0.03 ± 0.00	0.03 ± 0.00
NL	5	3.39 ± 0.02	3.40 ± 0.03	3.39 ± 0.02	3.38 ± 0.02	3.37 ± 0.05	3.38 ± 0.01
WW	5	0.22 ± 0.02	0.25 ± 0.01	0.24 ± 0.01	0.24 ± 0.02	0.26 ± 0.01	0.23 ± 0.01
DW	5	0.03 ± 0.00	0.03 ± 0.01	0.04 ± 0.00	0.04 ± 0.00	0.04 ± 0.02	0.04 ± 0.00
NL	9	3.85 ± 0.03	3.80 ± 0.03	3.82 ± 0.04	3.76 ± 0.07	3.74 ± 0.06	3.79 ± 0.06
WW	9	0.40 ± 0.01	0.38 ± 0.01	0.42 ± 0.03	0.37 ± 0.02	0.38 ± 0.01	0.41 ± 0.02
DW	9	0.03 ± 0.01	0.04 ± 0.01	0.04 ± 0.00	0.03 ± 0.01	0.03 ± 0.00	0.03 ± 0.00
NL	12	4.19 ± 0.13	4.19 ± 0.07	4.24 ± 0.03	4.25 ± 0.04	4.20 ± 0.07	4.28 ± 0.10
WW	12	0.63 ± 0.05	0.58 ± 0.05	0.55 ± 0.05	0.55 ± 0.05	0.53 ± 0.04	0.62 ± 0.04
DW	12	0.08 ± 0.01	0.07 ± 0.01	0.05 ± 0.01	0.07 ± 0.01	0.07 ± 0.01	0.07 ± 0.01
NL	16	4.96 ± 0.10	5.02 ± 0.05	4.92 ± 0.07	5.04 ± 0.06	5.07 ± 0.12	4.96 ± 0.22
WW	16	1.10 ± 0.17	1.20 ± 0.07	1.14 ± 0.09	1.17 ± 0.09	1.21 ± 0.08	1.22 ± 0.14
DW	16	0.12 ± 0.01	0.14 ± 0.01	0.13 ± 0.01	0.14 ± 0.01	0.14 ± 0.01	0.14 ± 0.02
NL	19	5.67 ± 0.11	5.71 ± 0.13	5.98 ± 0.10	5.69 ± 0.10	5.82 ± 0.22	5.87 ± 0.14
WW	19	1.99 ± 0.06	1.82 ± 0.12	2.01± 0.06	1.17 ± 0.12	1.99 ± 0.2	2.0 ± 0.18
DW	19	0.25 ± 0.01	0.23 ± 0.01	0.26 ± 0.01	0.22 ± 0.01	0.25 ± 0.02	0.23 ± 0.01
NL	26	6.70 ± 0.15	6.75 ± 0.17	6.84 ± 0.28	6.82 ± 0.09	6.76 ± 0.16	6.73 ± 0.18
WW	26	3.53 ± 0.18	4.07± 0.12	4.27 ± 0.44	3.80 ± 0.15	3.59 ± 0.22	3.94 ± 0.20
DW	26	0.59 ± 0.17	0.51± 0.02	0.54 ± 0.07	0.46 ± 0.02	0.45 ± 0.01	0.49 ± 0.05

DISCUSSION

Fatty Acid Composition of Enriched Rotifers

Rotifers enriched with 10% DHA contained a mean of 10.5% TFA DHA, while the unenriched rotifers contained no detectable DHA, consistent with the lack of DHA in *N. oculata*. A number of workers have reported a direct relationship between PUFA levels in rotifers and in their emulsion diets (Watanabe et al., 1983; Rainuzzo, Olson and Rosenlund 1989; Frolov, Pankov,

Geradze, Pankova and Spektorova 1991; Rezek, Watanabe, Harel, Seaton 2010; Carrier et al., 2011). Rotifers were cultured on *N. oculata*, which naturally contains ARA (3.94% TFA) (Reed Mariculture, Campbell, CA, US), causing the levels of ARA in rotifers from all ARA treatments to be slightly higher than emulsion levels, increasing from 1.5 to 8.2% TFA at treatment levels increasing from 0 to 6% ARA. The unenriched rotifers, on the other hand, contained an ARA level of 3.5% TFA, similar to their diet of *N. oculata* (3.94% TFA). Rotifers enriched with 10% DHA and 0% ARA had a lower ARA level (1.5%) than unenriched rotifers fed only *N. oculata* (3.5%), consistent with a reduction of relative ARA concentrations by enrichment with an emulsion devoid of ARA.

Table 5. Survival (mean ± SEM, *N* = 5) of larvae fed enriched rotifers *Brachionus rotundiformis* and *Artemia*, using 5 different oil emulsions consisting of 10% DHA and varying levels of ARA (0%, 1%, 2%, 4%, 6%) and an unenriched group. No significant treatment differences were observed for any parameter on any sampling date

Dph	10:0	10:1	10:2	10:4	10:6	Unenriched
2	80.1 ± 8.6	66.2 ± 8.4	63.6 ± 3.3	71.4 ± 7.5	78.2 ±4.0	70.0 ± 7.6
5	76.1 ± 7.7	58.9 ± 5.1	63.0 ± 2.9	70.6 ± 7.5	71.5 ± 4.1	69.7 ± 7.7
9	70.3 ± 7.9	57.2 ± 5.4	62.0 ± 2.5	68.9 ± 8.1	64.5 ± 4.2	69.5 ± 7.8
12	63.9 ± 6.9	55.4 ± 5.7	63.3 ± 3.2	60.1 ± 7.5	55.1 ± 2.5	58.3 ± 2.8
16	56.9 ± 8.4	49.6 ± 5.3	51.4 ± 8.1	58.5 ± 8.3	50.1 ± 3.3	52.4 ± 6.2
19	54.0 ± 7.8	48.9 ± 5.1	50.3 ± 8.1	54.2 ± 7.7	46.7 ± 4.3	49.0 ± 7.7
26	36.0 ± 8.6	39.0± 4.5	50.1 ± 9.2	42.3 ± 3.9	39.1 ± 5.9	44.9 ± 5.9

Table 6. Hypersaline stress resistance (ST-50) (mean ± SEM, *N* = 5) of larvae fed enriched rotifers *Brachionus rotundiformis* and *Artemia*, using 5 different oil emulsions consisting of 10% DHA and varying levels of ARA (0%, 1%, 2%, 4%, 6%) and an unenriched group. Larvae were subjected to a hypersaline stress test of 60 g L^{-1}. No significant treatment differences were observed

dph	10:0	10:1	10:2	10:4	10:6	Unenriched
13	12.1 ± 2.77	10.8 ± 0.89	13.6 ± 5.65	8.52 ± 1.69	20.1 ± 4.97	10.7 ± 1.68
27	37.7 ± 7.36	45.5 ± 9.23	31.7 ± 4.58	40.0 ± 4.33	47.3 ± 8.45	38.9 ± 7.30

Mean EPA was 7.9% TFA in rotifers fed the enrichment emulsions and 15.3% TFA in the unenriched rotifers. While EPA was not supplied in the enrichment emulsions, rotifers were cultured before enrichment on *N. oculata,* which contains 31.4% EPA, which contributed to substantial EPA in enriched rotifers.

The lower than expected amount of EPA (15.3% TFA) in the unenriched rotifers relative to their diet of *N. oculata* could be a result of the rotifers converting EPA (20:5n-3) to DPA (22:5n-3), which averaged 6.3% TFA, a significantly higher amount than in rotifers fed the enrichment emulsions (2.9% DPA TFA).

The presence of EPA in the emulsion-enriched rotifers (originating from *N. oculata*) resulted in an average DHA/EPA ratio of 1.4, which is lower than the recommended minimum of 2:1 DHA: EPA for marine finfish larvae (Tocher and Sargent 1984; Sargent et al., 1999).

Fatty Acid Composition of Enriched *Artemia*

In this study, *Artemia* enriched with 10% DHA contained a mean of only 7.3% DHA TFA, while the unenriched *Artemia* contained no detectable DHA. Compared to rotifers, *Artemia* enrichment is difficult due to their rapid development and catabolism of DHA (Sorgeloos, Dhert and Candreva 2001). It is possible that higher levels of DHA in *Artemia* could be achieved if the enrichment period is extended from 16 h (as used in this study) to 19-24 h (McEvoy, Navarro, Bell and Sargent 1995).

Enrichment success in *Artemia* also depends on genetic characteristics. Chinese strains of *Artemia* accumulated higher levels of DHA than the Great Salt Lake strain or the San Francisco Bay strain used in this study (Dhert, Sorgeloos and Devresse 1993; Evjemo, Coutteau, Olsen and Sorgeloos 1997).

The higher (1.7) DHA/EPA ratio in *Artemia* is attributed to lower EPA levels, which averaged 3.7% TFA in *Artemia* enriched with emulsions and 5.5% TFA in unenriched *Artemia* nauplii. ARA levels in enriched *Artemia* were slightly higher than supplied in the prescribed enrichments, except in the 6% ARA treatment which contained 5.2% ARA.

The higher than prescribed ARA levels are likely attributable to endogenous ARA (0.8% TFA) in *Artemia* (Coutteau and Mourente 1997). *Artemia* fed 0% ARA had a lower ARA level (1.5%) than unenriched *Artemia* fed only *N. oculata* (3.5%), again consistent with a reduction of relative ARA concentrations by enrichment with an emulsion devoid of ARA.

Fatty Acid Composition of D27ph Larvae

On d27ph, mean DHA levels in larvae fed prey enriched with 10% DHA was 9.8% TFA, corresponding to the levels provided during the rotifer feeding stage. The presence of DHA (2.6% TFA) in southern flounder larvae fed unenriched prey, which contained no detectable DHA, is similar to what was found in larval turbot (Estevez, McEvoy, Bell, Sargent 1999), yellowtail flounder *Limanda ferruginea* (Storer 1839) (Copeman et al., 2002), summer flounder (Willey, Bengtson, Harel, 2003) and Senegal sole *Solea senegalensis* (Kaup 1858) (Villalta, Estevez, Bransden, Bell 2005, Morais, Narciso, Dores, and Pousao-Ferreira 2004) when fed prey containing no DHA. As suggested by previous studies (Morais et al., 2004), it is possible that these larvae fed 0% DHA derived a significant amount of DHA from endogenous yolk reserves, since marine fish eggs normally contain high concentration of PUFA's, especially DHA (Tocher, Fraser and Sargent 1985). Furthermore, DHA levels of 26.1% have been reported in high quality southern flounder eggs (Woolridge 2005). Alternatively, DHA may have been synthesized from precursor fatty acids 18:3n3, 20:5n3 or 22:5n3, which were at significantly higher concentrations in the unenriched rotifers and *Artemia* compared to prey enriched with emulsions. Conversion of 18:3n3 and 20:5n3 to DHA, while limited in most marine carnivorous fish species, has also been reported in the juvenile turbot (Linares and Henderson 1991; Mourente et al., 1991) and gilthead seabream (Mourente, Rodriguez and Tocher 1993).

Carcass ARA levels on d27ph were higher than levels provided in enriched rotifers and *Artemia*. This suggests that dietary ARA was accumulated in larval tissues to levels above those provided in their diet, as reported in larval turbot (Linares and Henderson 1991) and mahi *Coryphaena hippurus* (Linnaues 1758) (Ako, Kraul and Tamaru 1991). As reported in gilthead seabream larvae (Bessonart et al., 1999), tissue levels of ARA in southern flounder larvae increased with increasing dietary levels, suggesting that larval requirements for ARA can be met through dietary manipulations.

Growth and Survival

In this study, growth in terms of notochord length and wet and dry weight was not different in larvae fed prey enriched with 10% DHA from those fed unenriched prey with 0% DHA. This is in contrast to what was reported in red sea bream *Pagrus major* (Temmink and Shlegel 1843), where larvae fed rotifers enriched with high levels of DHA (≥ 20%) grew faster than those fed rotifers enriched with ≤ 2% DHA (Watanabe et al., 1989). Improved growth with higher dietary DHA has also been reported in larval yellowtail *Seriola quinqueradiata* (Temmink and Shlegel 1845) (Furuita, Takeuchi, Watanabe, Fujimoto, Sekiya and Imaizumi 1996, Copeman et al., 2002), yellowtail snapper *Ocyurus chrysurus* (Block 1791) (Faulk and Holt 2005a), cobia *Rachycentron canadum* (Linnaeus 1776) (Faulk and Holt 2005b), and black sea bass *Centropristis striata* (Linnaeus 1758) (Rezek et al., 2010). As previously reported for other flatfish species, including Senegal sole (Morais et al., 2004, Villalta et al., 2005) and Japanese flounder *Paralichthys olivaceus* (Temminck and Schlegel 1843) (Izquierdo, Arakawa, Takeuchi, Haroun, Watanabe 1992), a diet devoid of DHA was sufficient for southern flounder larval growth. These flatfish species apparently derive a significant amount of DHA from endogenous yolk reserves, or from elongation of precursor fatty acids 18:3n3, 20:5n3 or 22:5n3 (Mourente et al., 1993, Linares and Henderson 1991; Mourente et al., 1991).

Southern flounder larvae fed diets containing 0 – 6% ARA showed no significant differences in growth up to d26ph. Similar findings were reported for other species such as Japanese flounder (Estevez, Kaneko, Seikai, Tagawa and Tanaka 2001), larval turbot (Estevez, et al., 1999) and larval gilthead seabream (Koven et al., 2001) when fed diets containing 0 - 15% ARA. In larval white bass *Morone chrysops* (Rafinesque 1820), dietary ARA levels of 0 to 26% TFA also had no effect on growth, suggesting bioconversion of 18:2n-6 linoleic acid (LA) to ARA at sufficient levels for larval performance (Harel and Place 2003). From the results of the present study in southern flounder, it appears that the levels of dietary ARA originating from *N. oculata* fed to rotifers and those naturally found in *Artemia* instar 1 nauplii are sufficient for the growth of larval southern flounder.

In this study, *N. oculata* was added daily as background algae (greenwater) to the larval rearing tanks, a standard practice for larval rearing of southern flounder (Daniels and Watanabe 2003), as well as for many marine finfish species (Reitan, Rainuzzo, Øie and Olsen 1997; Faulk and Holt 2005b). *N. oculata* in the rearing tanks likely helped maintain the nutritional value of prey to the fish larvae. While larval southern flounder are considered strict carnivores, incidental ingestion of *N. oculata* cells by larvae is possible and may have contributed to dietary ARA in this study.

In agreement with the growth data, there were no significant differences in survival of southern flounder larvae fed diets containing 0 – 6% ARA up to d26ph. Larval turbot fed prey containing from 3 to 15% ARA also showed no significant differences in survival to d43ph (Estevez et al., 1999). In larval summer flounder, dietary ARA within the range of 0% to 12% TFA had no effect on growth or survival to d23ph (Willey et al., 2003).

Contrary to these findings, black seabass larvae fed prey containing 10% DHA, with increasing ARA within the range of 0-6%, showed improved growth and survival from first feeding through metamorphic stages (Rezek et al., 2010) when grown under greenwater conditions using *N. oculata* as a background microalgal species. These results emphasize species-specific differences in ARA requirements of marine finfish larvae.

There were no significant differences among treatments in the ability for southern flounder larvae to survive an acute hypersaline challenge of 60 g L^{-1}. Mean ST-50 increased from 12.6 min among treatments on d13ph to 40.2 min on d27ph, suggesting that resistance of larvae to hypersalinity stress increased during larval development (Bodinier, Sucre, Lecurieux-Belfond, Blondeau-Bidet and Charmantier 2010; Watanabe, Kuo and Huang 1985). The use of ARA supplementation during the early larval stages improved survival of larvae through post-metamorphic stages in gilthead seabream and summer flounder (Koven et al., 2001; Willey et al., 2003). Hence, it is possible that ARA effects would be manifested if survival were monitored through later post-metamorphic stages.

Interestingly, low levels of ARA were detected in the 0% ARA enriched rotifers (1.5% TFA) and unenriched *Artemia* nauplii (1.0% TFA) suggesting that even small amounts of ARA in the diet of southern flounder larvae were sufficient for normal stress resistance, growth and survival. Larval summer flounder fed a diet containing 6% ARA (TFA) showed increased hypersaline stress resistance when exposed to 70 gL^{-1} on d18ph and to 80 gL^{-1} on d45ph (Willey et al., 2003). Black sea bass larvae fed diets supplemented with 6-12% ARA demonstrated a significant increase in Na^+/K^+ ATPase mRNA 24 h after a sublethal salinity (42 g L^{-1}) challenge compared to larvae fed 0% ARA (Carrier et al., 2011). It is possible the acute hypersaline challenge (60 g L^{-1}) used in the present study overwhelmed the larvae and did not allow the effects of dietary ARA supplementation on osmoregulation to be clearly resolved. Lower or fluctuating salinity challenge, analysis of Na^+ K^+ ATPase gene expression (Carrier et al., 2011, Ostrowski, Watanabe, Montgomery, Rezek, Shafer and Morris 2011), or cortisol production analysis may enable better resolution of the effects of dietary DHA or ARA on resistance to salinity stress (Jensen, Madsen and Kristiansen 1998; Koven, 2003; Van Anholt, Koven and Lutzky 2004).

In summary, dietary DHA levels (0-10%) combined with dietary ARA within the range of 0 to 6% had no significant effects on growth, survival and hypersaline stress resistance in larval southern flounder through metamorphic stages. Southern flounder larvae were apparently able to accumulate sufficient ARA from their live prey even though present in these organisms in seemingly small concentrations. Carcass DHA levels in d27ph larvae fed live prey lacking this essential fatty acid indicated that a significant amount of DHA was derived from endogenous yolk reserves, or that this species can synthesize DHA from 18 carbon and 20 carbon n3 precursors at levels that are sufficient for normal growth and survival.

ACKNOWLEDGMENTS

This study was supported by a grant from the United States Department of Agriculture, National Institute of Food and Agriculture (USDA-NIFA) and from Marine Biotechnology in North Carolina (MARBIONC). We thank Patrick Carroll, Kimberly Copeland and Scott Wheatley for technical support.

REFERENCES

Ackman, R. G. (1979) Fish lipids. Part 1. Papers presented at the Jubilee Conference of the Torry Research Station Aberdeen, Scotland. *Advances in Fish Science and Technology* 86-103.

Ako, H., Kraul, S. and Tamaru, C. (1991) Pattern of fatty acid loss in several warmwater fish species during early development. Larvi 91 Fish and Crustacean Larivculture Symposium. *European Aquaculture Society, Special Publication No. 15.*

Bell, M. V., Henderson, R. J., Pirie, B. J. S., and Sargent, J. R. (1985) Effects of dietary polyunsaturated fatty acids deficiencies on mortality, growth and gill structure in the turbot, *Scopthalmus maximus. Journal of Fish Biology* 26, 181-191.

Berlinsky, D. L., King, W. V., Smith, T. I. J., Hamilton II, R. D., Holloway, J., Jr., and Sullivan, C. V. (1996) Induced ovulation of southern flounder *Paralichthys lethostigma* using gonadotropin releasing hormone analogue implants. *Journal of the World Aquaculture Society* 27, 143-152.

Bessonart, M., Izquierdo, M. S., Salhi, M., Hernandez-Cruz, C. M., Gonzalez, M. M., and Fernandez-Palacios, H. (1999) Effect of dietary arachidonic acid levels on growth and survival of gilthead seabream *Sparus aurata* L. larvae. *Aquaculture* 179, 265-275.

Bodinier, C., Sucre, E., Lecurieux-Belfond, L., Blondeau-Bidet, E., and Charmantier, G. (2010) Ontogeny of osmoregulation and salinity tolerance in the gilthead sea bream *Sparus aurata. Comparative Biochemistry and Physiology Part A* 157, 220-228.

Carrier, J. K. III, Watanabe, W. O., Harel, M., Rezek, T. C., Seaton, P. J., and Shafer, T. H. (2011) Effects of dietary arachidonic acid on larval performance, fatty acid profiles, stress resistance, and expression of Na^+/K^+ ATPase mRNA in black sea bass *Centropristis striata. Aquaculture* 319, 111-121.

Copeman, L. A., Parrish, C. C., Brown, J. A., and Harel, M. (2002) Effects of docosahexanoic, eicosapentaenoic, and arachidonic acids on the early growth, survival, lipid composition and pigmentation of yellowtail flounder *Limanda ferruginea*: a live food enrichment experiment. *Aquaculture* 210, 285-304.

Coutteau, P. and Mourente, G. (1997) Lipid classes and their content of n-3 highly unsaturated fatty acids (HUFA) in *Artemia franciscana* after hatching, HUFA-enrichment and subsequent starvation. *Marine Biology* 130, 81-91.

Daniels, H. V. and W. O. Watanabe. 2003. A practical hatchery manual: Production of southern flounder fingerlings. *North Carolina Sea Grant*. 40 pp.

Dhert, P., Lavens, P. and Sorgeloos, P. (1991) Stress evaluation: a tool for quality control of hatchery produced shrimp and fish fry. Laboratory of Aquaculture and Artemia Reference Center. *Rozier* 44, B-9000 Gent, Belgium.

Dhert, P., Sorgeloos, P. and Devresse, B. (1993) Contributions towards a specific DHA enrichment in the live food *Brachionus plicatilis* and *Artemia* sp. In: H. Reinertsen, L. A. Dahle, L. Jorgenson, K. Tvinnerein (Eds.), *Fish Farming Technology: Proceedings of the 1st International Conference on Fish Farming Technology*, Trondheim, Norway, 9-12 August 1993. AA Balkema, Rotterdam, pp. 109-114.

Estevez, A., McEvoy, L. A., Bell, J. G., and Sargent, J. R. (1999) Growth, survival, lipid composition and pigmentation of turbot *Schopthalmus maximus* larvae fed live prey enriched in arachidonic and eicosapentaenoic acids. *Aquaculture* 180, 321-343.

Estevez, A., Kaneko, T., Seikai, T., Tagawa, M., Tanaka, M. (2001) ACTH and MSH production in Japanese flounder *Paralichthys olivaceus* larvae fed arachidonic acid enriched live prey. *Aquaculture* 192, 309-319.

Evjemo, J. O., Coutteau, P., Olsen, Y., and Sorgeloos, P. (1997) The stability of docosahexaenoic acid in two *Artemia* species following enrichment and subsequent starvation. *Aquaculture* 155, 135-148.

Faulk, C. K. and Holt, J. G. (2005a). Evaluation of fatty acid enrichment of live food for yellow tail snapper *Ocyurus chrysurus* larvae. *Journal of the World Aquaculture Society* 36, 271-281.

Faulk, C. K., Holt, J. G. (2005b). Advances in rearing cobia *Rachycentron canadum* larvae in recirculating aquaculture systems: Live prey enrichment and greenwater culture. *Aquaculture* 249, 231-243.

Folch, J., Lees, M. and Sloane Stanley, G. H. (1957) A simple method for the isolation and purification of total lipids from animal tissues. *Journal of Biological Chemistry* 226, 497-509.

Frolov, A. V., Pankov, S. L., Geradze, K. N., Pankova, S. A., and Spektorova, L. V. (1991) Influence of the biochemical composition of food on the biochemical compostition of the rotifer *Brachionus plicatilis*. *Aquaculture* 97, 181-202.

Furuita, H., Takeuchi, T., Watanabe, T., Fujimoto, H., Sekiya, S., and Imaizumi, K. (1996) Requirements of larval yellowtail for eicosapentaenoic acid, docosahexaenoic acid, and highly unsaturated fatty acid. *Fisheries Science* 62(3), 372-379.

Harel, M. and Place, A. R. (2003) Tissue essential fatty acid compostion and competitive response to dietary manipulations in white bass *Morone chrysops*, striped bass (*M. saxatilis*) and hybrid striped bass *M. chrysops x M. saxatilis*. *Comparative Biochemistry and Physiology Part B* 135, 83-94.

Ibeas, C., Cejas, J. R., Fores, R., Badia, P., Gomes, T., Hernandez, A. L. (1997) Influence of eicosapentaenoic to docosahexaenoic acid ratio (EPA/DHA) of dietary lipids on growth and fatty acid composition of gilthead seabream *Sparus aurata* juveniles. *Aquaculture* 150, 91-102.

Izquierdo, M. S., Arakawa, T., Takeuchi, T., Haroun, R., and Watanabe, T. (1992) Effect on n-3 HUFA levels in *Artemia* on growth of larval Japanese flounder *Paralichthys olivaceus*. *Aquaculture* 105, 73-82.

Jensen, M. K., Madsen, S. S. and Kristiansen, K. (1998) Osmoregulation and salinity effects on the expression and activity of Na^+ K^+-ATPase in gills of European Sea Bass *Dicentrarchus labrax* L. *The Journal of Experimental Zoology* 282, 290-300.

Koven, W. M., Kissil, G. W. and Tandler, A. (1989) Lipid and n-3 requirement of *Sparus aurata* larvae during starvation and feeding. *Aquaculture* 79, 185-192.

Koven, W., Barr, Y., Lutzky, S., Ben-Atia, I., Weiss, R., Harel, M., Behrens, P., and Tandler, A. (2001) The effect of dietary arachidonic acid (20:4n-6) on growth, survival and resistance to handling stress in gilthead seabream *Sparus aurata* larvae. *Aquaculture* 193, 107-122.

Koven, W. M. (2003) Key factors influencing juvenile quality in mariculture. *Israeli Journal of Aquaculture Bamidgeh* 55(4), 283-297.

Linares, F. and Henderson, R. J. (1991) Incorporation of ^{14}C-labelled polyunsaturated fatty acids by juvenile turbot, *Scophthalmus maximus* L. *in vivo*. *Journal of Fish Biology* 38, 335-347.

Mangino Jr., A. and Watanabe, W. O. (2006) Combined effects of turbulence and salinity on growth, survival and whole body osmolality of larval southern flounder. *Journal of the World Aquaculture Society* 37, 407-420.

McEvoy, L., Navarro, J. C., Bell, J. G., and Sargent, J. R. (1995) Autoxidation of oil emulsions during the *Artemia* enrichment process. *Aquaculture* 134, 101-112.

Morais, S., Narciso, L. and Pousao-Ferreira, P. (2004) Lipid enrichment for Senegalese sole *Solea senegalensis* larvae: effect on larval growth, survival and fatty acid profile. *Aquaculture International* 12, 281-298.

Mourente, G., Tocher, D. R. and Sargent, J. R. (1991) Specific accumulation of docosahexaenoic acid (22:6n-3) in brain lipids during development of juvenile turbot *Scophthalmus maximus* L. *Lipids* 26, 871-877.

Mourente, G., Rodriguez, A., Tocher, D. R., and Sargent, J. R. (1993) Effects of dietary docosahexaenoic acid (22:6n-3) on lipid and fatty acid compositions and growth in gilthead sea bream (*Sparus aurata* L.) larvae during the first feeding. *Aquaculture* 112, 79-98.

Moustakas, C., Watanabe, W. and Copeland, K. (2004) Combined effects of photoperiod and salinity on growth, survival, and osmoregulatory ability of larval southern flounder *Paralichthys lethostigma*. *Aquaculture* 229, 159-179.

Ostrowski, A., Watanabe, W., Montgomery, F., Rezek, T., Shafer, T., and Morris, J. (2011) Effects of salinity and termperature on growth, survival, whole body osmolality, and expression of Na^+ / K^+ ATPase mRNA in red porgy (*Pagrus pagrus*) larvae. *Aquaculture* 314, 193-201.

Rainuzzo, J. R., Olsen, Y. and Rosenlund, G. (1989) The effect of enrichment diets on the fatty acids composition of the rotifer *Brachionus placatilis*. *Aquaculture* 79, 157-161.

Rainuzzo, J. R., Reitan, K. I. and Olsen, Y. (1997) The significance of lipids at early stages of marine fish: a review. *Aquaculture* 155, 103-115.

Reitan, K. I., Rainuzzzo, J. R., Øie, G., and Olsen, Y. (1997) A review of the nutritional effects of algae in marine fish larvae. *Aquaculture* 155: 207-221.

Rezek, T. C., Watanabe, W. O., Harel, M., and Seaton, P. J. (2010) Effects of dietary docosahexaenoic acid (22:6n-3) and arachidonic acid (20:4n-6) on the growth, survival, stress resistance and fatty acid composition in black sea bass *Centropristis striata* (Linnaeus 1758) larvae. *Aquaculture Research* 41: 1302-1314.

Sargent, J. R., Bell, G., McEvoy, L., Tocher, D., and Estevez, A. (1999) Recent developments in the essential fatty acid nutrition of fish. *Aquaculture* 177, 191-199.

Smith, T. J., McVey, D. and Wallace, J. (1999) Broodstock management and spawning of Southern flounder *Paralichthys lethostigma*. *Aquaculture* 176, 97-99.

Sorgeloos, P., Dhert, P. and Candreva, P. (2001) Use of brine shrimp, *Artemia* spp., in marine fish larviculture. *Aquaculture* 200, 147-159.

Tacon, A. G. (1990) *Standard methods for the nutrition and feeding of farmed fish and shrimp.* Argent Laboratories Press, Redmond, Washington, US.

Takeuchi, T., Toyota, S., Satoh, Watanabe, T. (1990) Requirement of juvenile red sea bream *Pagrus major* for eicosapentaenoic and docosahexaenoic acids. *Nippon Suisan Gakkaishi* 56, 1263-1269.

Takeuchi, T., Masuda, R., Ishizaki, Y., Watanabe, T., Kanematsu, M., Imaizumi, K., Tsukamoto, K. (1996) Determination of the requirement of larval striped jack for eicosapentaenoic acid and docosahexaenoic acid using enriched *Artemia* nauplii. *Fisheries Science* 62(5), 760-765.

Tocher, D. R. and Sargent, J. R. (1984) Analysis of lipids and fatty acids in ripe roes of some Northwest European marine fish. *Lipids* 7, 492-499.

Tocher, D. R., Fraser, A. J. and Sargent, J. R. (1985) Fatty acid composition of phospholipids and neutral lipids during embryonic and early larval development in Atlantic herring *Clupea harengus*. *Lipids* 20, 69-74.

Van Anholt, R. D., Koven, W. M., Lutzky, S., and Wendelaar Bonga, S. E. (2004) Dietary supplementation with arachidonic acid alters the stress response of gilthead seabream (*Sparus aurata*) larvae. *Aquaculture* 238, 369-383.

Villalta, M., Estevez, A., Bransden, M. P., and Bell, J. G. (2005) The effect of graded concentrations of dietary DHA on growth survival and tissue fatty acid profile of Senegal sole *Solea senegalensis* larvae during the Artemia feeding period. *Aquaculture* 249 353-365.

Watanabe, T., Kitajima, C. and Fujita, S. (1983) Nutritional values of live organisms used in Japan for mass propagation of fish: a review. *Aquaculture* 34, 115-143.

Watanabe, T., Izquierdo, M. S., Takeuchi, T., Satoh, S., and Kitajima, C. (1989) Comparison between eicosapentaenoic and docosahexaenoic acids in terms of essential fatty acid efficacy in larval red seabream. *Nippon Suissan Gakkaishi* 55(9), 1635-1640.

Watanabe, T. (1993) Importance of docosahexaenoic acid in marine larval fish. *Journal of the World Aquaculture Society* 24, 152-161.

Watanabe, W. O., Kuo, C. M. and Huang, M. C. (1985) The ontogeny of salinity tolerance in the tilapias *Oreochromis aureus*, *O. niloticus* and an *O. mossambicus* x *O. niloticus* hybrid, spawned and reared in freshwater. *Aquaculture* 47, 353-367.

Watanabe, W. O., Carroll, P. M. and Daniels, H. V. (2001) Sustained, natural spawning of Southern Flounder *Paralichthys lethostigma* under an extended photothermal regime. *Journal of World Aquaculture Society* 32, 153-166.

Watanabe, W. O., Woolridge, C. A. and Daniels, H. V. (2006) Progress toward year-round spawning of southern flounder broodstock by manipulation of temperature and photoperiod. *Journal of the World Aquaculture Society* 37, 256-272.

Willey, S., Bengtson, D. A. and Harel, M. (2003) Arachidonic acid requirements in larval summer flounder, *Paralichthys dentatus*. *Aquaculture International* 11, 131-149.

Woolridge, C. (2005) Phosphorus composition and fatty acid profiles as determinants of egg quality in southern flounder *Paralichthys lethostigma*. *M.S. Thesis*, Department of Biology and Marine Biology, University of North Carolina Wilmington.

In: Arachidonic Acid
Editor: Jason M. O'Keefe
ISBN: 978-1-63117-619-7

Chapter 2

DIETARY SOURCES AND HEALTH IMPLICATIONS OF ARACHIDONIC ACID

Susana Vargas Martins[1], Cristina Mateus Alfaia[1], Paula Alexandra Lopes[1] and José Mestre Prates[1,*]

[1]CIISA, Faculdade de Medicina Veterinária,
Universidade de Lisboa,
Pólo Universitário do Alto da Ajuda,
Lisboa, Portugal

ABSTRACT

Dietary habits in Western societies, in particular the amount and type of fat consumed, play an important role in human health condition. The present chapter reviews the biosynthesis, dietary sources and health implications of the arachidonic acid (20:4*n*-6, AA), a long-chain polyunsaturated fatty acid (PUFA) from the *n*-6 family found in mammalian cells, especially in membrane phospholipids. Systemic and tissue levels of AA may result from two origins, the endogenous synthesis from linoleic acid (18:2*n*-6, LA) elongation and the exogenous dietary source. LA elongation involves desaturation and chain elongation enzymes, which are also common to the *n*-3 PUFA α-linolenic acid (18:3*n*-3, ALA) elongation pathway. As dietary intake of ALA is low, the production of AA from LA is little restrained. Humans consume AA from animal-derived foods (e.g., eggs and meat) as it does not occur in plants.

* E-mail: japrates@fmv.ulisboa.pt.

The major biological functions of AA are as cell signaling molecule and as precursor to an extensive group of compounds, called eicosanoids. Different eicosanoids, namely prostaglandins, thromboxanes, leukotrienes, and other bioactive metabolites, are produced by different types of cells via distinct enzymatic pathways. These AA mediators act like potent chemical messengers on immunity regulation. The overproduction of prothrombotic and pro-inflammatory eicosanoids, thromboxanes and leukotrienes derived from AA, has been associated with numerous diseases, such as cancer, heart disease, rheumatoid arthritis or asthma. In addition, AA and its end-products are involved in physiological processes in the central nervous system and have been implicated in neurological, neurodegenerative, and psychiatric disorders. During several decades, AA and its metabolites have been considered nothing but pro-inflammatory with recommendations to lower its intake from the diet. However, the discovery of anti-inflammatory mediators produced from AA is changing the long-held misunderstandings about this fatty acid. Unlike the pro-inflammatory prostaglandins and leukotrienes, AA-derived lipoxins display counter-regulatory actions by promoting the resolution of acute inflammatory responses. In conclusion, AA has been the focus of new investigations in the recent years and, rather than being detrimental, moderate AA intake seems to be a key dietary component for cardiovascular and cognitive health, with harmless effects in healthy subjects.

1. INTRODUCTION

Lipids were originally regarded as just membrane constituents and energy storing molecules but are now recognized as potent signaling molecules that regulate a multitude of cellular responses via receptor-mediated pathways, including cell growth and death, or inflammation and infection. Araquidonic acid (20:4*n*-6, AA) is the most prevalent long-chain polyunsaturated fatty acid (LC-PUFA) from the *n*-6 family. While is a basic constituent of cells, especially in membrane phospholipids, AA is the main eicosanoid precursor displaying unique properties, thus making its role in inflammation distinct from that of other lipids derived from the same PUFA. Although it is well established that the pro-inflammatory mediators generated in the inflammed tissue drive acute inflammation, there is also systemic and local production of endogenous mediators that counter-balance these pro-inflammatory events, evolved to avoid the development of several pathologies. Lipid mediators derived from AA, and also from the *n*-3 PUFA, eicosapentaenoic (20:5*n*-3, EPA) and docosahexaenoic (22:6*n*-3, DHA) acids, are synthesized during normal cell homeostasis or, more often, after cell activation and in conditions of stress, functioning as activators of counter-regulatory, anti-inflammatory and proresolution mechanisms. The diversity of their actions arises because such metabolites are produced via discrete enzymatic pathways and because it elicits their response via different receptors. Interestingly, these immune-modulatory effects are also found in a family of lipids, prostanoids, which are well-appreciated to be involved in driving some of the cardinal signs of inflammation (heat, redness, swelling, pain and loss of function).

As the role of lipids in inflammation is diverse, this review provides an update on the bioactive lipid research on AA to date and summarizes the main findings on its metabolism and biological functions, dietary sources and nutritional aspects, besides the current controversy around its health implications. It will include signaling molecules derived from AA (namely prostanoids, leukotrienes, epoxyeicosatrienoic acids and lipoxins), and also from the *n*-3 PUFA, EPA (in particular, the E-series resolvins), and DHA (D-series resolvins, protectins and maresins), with contrasting metabolic effects. These lipid mediators not only drive acute inflammation but, in turn, can counter-regulate its severity and bring about its timely resolution across a wide range of XXI century pathologies, from cancer to cardiovascular diseases and metabolic syndrome; from neurodegenerative to autoimmune disorders.

2. Biosynthesis and Metabolism of Arachidonic Acid

AA (5,8,11,14-eicosatetraenoic acid) is a LC-PUFA from the *n*-6 family containing 20 carbons. In mammals, the precursors for the synthesis of LC-PUFA of *n*-6 and *n*-3 series are linoleic acid (18:2*n*-6, LA) and α-linolenic acid (18:3*n*-3, ALA), respectively. Due to the lack of adequate enzymes (Δ^{12} and Δ^{15} desaturases), LA and ALA cannot be synthesized *de novo* in animal cells and, therefore, must be supplied from the diet (Kolanowski, 2007, cited by Bederska-Lojewska et al., 2013). These essential fatty acids (LA and ALA), once absorbed by enterocytes, are mainly transported to the liver where undergo desaturation (increase in the number of double bonds) and/or elongation (increase in the number of the carbon atoms) through the same desaturase and elongase enzymes (Giudetti and Cagnazzo, 2012). The *n*-6 fatty acid LA can be converted into AA as the *n*-3 fatty acid ALA into EPA and DHA. However, excessive amounts of LA slow down the formation of EPA and DHA (Flachs et al., 2009). The initial rate limiting desaturation of LA to γ-linolenic acid (18:3*n*-6, GLA) is catalyzed by the Δ^6 desaturase enzyme (Stoffel et al., 2008). Then, elongation takes place to convert GLA to dihomo-γ-linolenic acid (20:3*n*-6, DGLA), by elongation of very long-chain fatty acids, and, finally, a cycle of elongation and desaturation by Δ^5 desaturase generates AA (Moon et al., 2009) (Figure 1).

AA is the major constituent of mammalian cell membranes stored in glycerophospholipids and esterified in the *sn*-2 position (Balsinde et al., 1995). It is present in high concentrations in nervous tissue, such as brain, synapses, myelin, astrocytes, and in kidneys and heart (Wainwright, 2002). AA is also present at relatively high levels in cell membranes of erythrocytes, neutrophils, monocytes, as well as, in hepatocytes (Simopoulos, 2001), whereas DHA is found at high levels in only few tissues outside the central nervous system (Bederska-Lojewska et al., 2013). AA in the diet can be efficiently absorbed and incorporated into tissue membranes. When tissues are exposed to multiple physiological and pathological stimuli, such as growth factors, neurotransmitters, and cytokines (Farooqui et al., 2006), AA is cleavaged and released from membrane phospholipids by the action of selective phospholipase enzymes (Harizi et al., 2008). Once released, free AA may be reincorporated into membrane phospholipids, diffused outside the cell or metabolized. The primary site for PUFA metabolism is the liver. However, it can also take place in other tissues (Hughes and Dhiman, 2002).

The production of free AA is a highly regulated process that represents a balance between two competing reactions, namely, phospholipid deacylation, by phospholipase enzymes and reacylation and transfer into various phospholipid pools, by acyltransferases and transacylases (Pérez-Chacón et al., 2009). The deacylation reaction of phospholipids may be mediated by three phospholipase enzymes, phospholipase A_2 (PLA_2), phospholipase C (PLC), and phospholipase D (PLD). However, only PLA_2, located in the endoplasmic reticulum or nuclear membrane, seems to be able to release free AA directly in a single-step reaction, by hydrolyzing an ester bond at the *sn*-2 position of membrane phospholipids (Brash, 2001). The activity of PLA_2 is thought to be the primary rate-limiting step controlling the supply of free AA for further metabolism in most cell types.

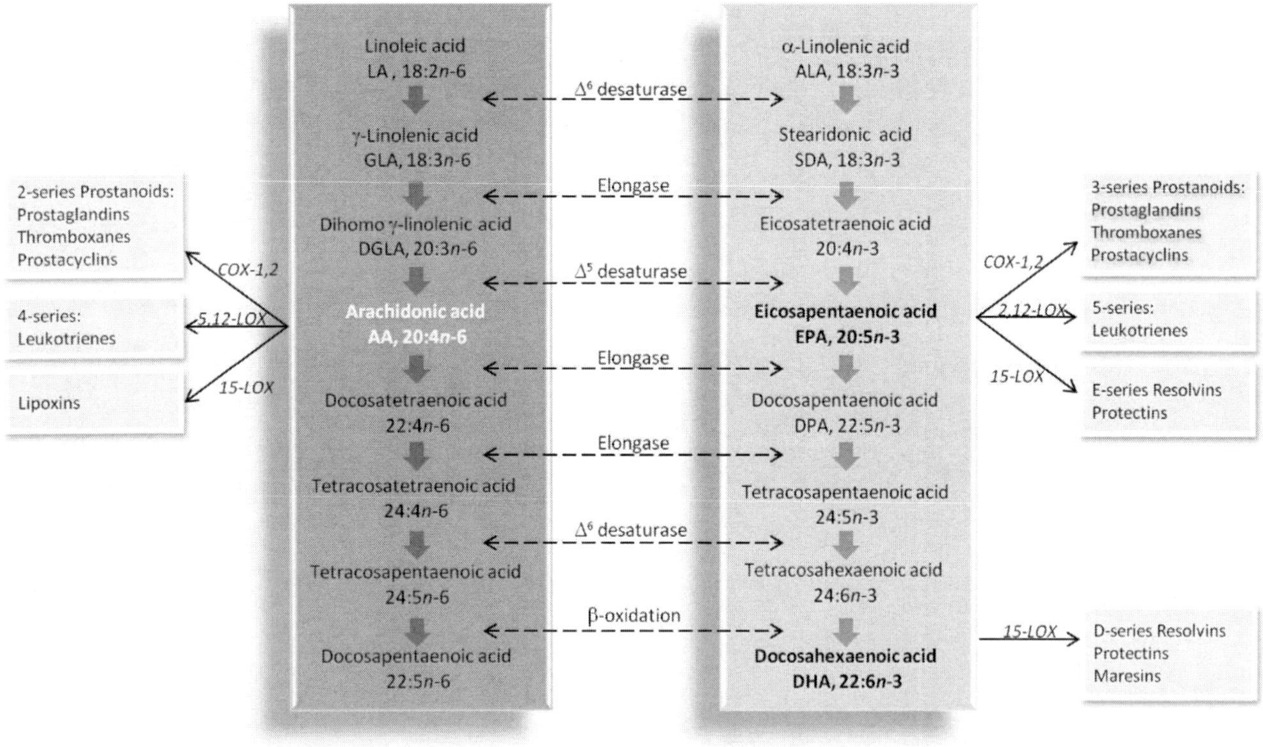

(Adapted from Bosma-den Boer et al., 2012).

Figure 1. Metabolic pathways of *n*-6 and *n*-3 PUFA.

PLA_2 forms a large family of enzymes, differently regulated, divided into four major classes: secretory and cytosolic calcium (Ca^{2+})-dependent, cytosolic and peroxisomal Ca^{2+}-independent, which vary in localization, structure, molecular weight, fatty acid chain specificity and Ca^{2+} dependence (Balsinde et al., 2002; Burke and Dennis, 2009). By contrast, PLC and PLD do not release free AA directly; rather, these enzymes generate lipid products containing AA (diacylglycerol and phosphatidic acid, respectively), which can be released subsequently by diacylglycerol- and monoacylglycerol-lipases (Pérez-Chacón et al., 2009).

Alternatively, AA may be metabolized into a family of pharmacologically active metabolites, known as eicosanoids with pivotal biological effects on intracellular targets. The binding of growth factors and hormones to membrane receptors, leads to activation of PLA_2 and release of AA which become a substrate for eicosanoid biosynthesis (Russo, 2009). However, under activation, AA reacylation is still very significant because only a minor fraction of the AA released by PLA_2 is available for eicosanoid synthesis. The remainder is incorporated back into phospholipids by acyltransferases in various cells, particularly those involved in inflammatory reactions, such as neutrophils and macrophages (Chilton et al., 1996). Upon release from membrane phospholipids, the AA synthesized from the DGLA desaturation by Δ^5 desaturase, can be further elongated to docosatetraenoic acid (22:4*n*-6) or to its respective set of eicosanoids metabolites (Patterson et al., 2012).

AA convertion may follow four principal pathways to produce an important family of oxygenated products. These metabolites derived from AA are produced through a series of complex interrelated biosynthetic pathways, sometimes designated arachidonate or eicosanoid cascade: i) the cyclooxygenase (COX) pathway, which acts via a series of G-protein-coupled receptors to generate prostaglandins (PG), prostacyclins, and thromboxanes (TX), collectively termed prostanoids; ii) the lipoxygenase (LOX) pathway, which produces leukotrienes (LT), and lipoxins (LX); iii) the non-enzymatic lipid peroxidation, which produces isoprostanes (stereoisomers of prostaglandins); iv) the cytochrome P450 (CYP) monooxygenase pathway, which comprises a family of LOX-like hydroeicosatetraenoic acids (HETE), epoxyeicosatrienoic acids (EET) and their corresponding dihydroxy derivatives (DHET) (Capdevila et al., 2000).

AA can be metabolized by COX, which has constitutive COX-1 and inducible COX-2 isoforms sharing 60% of amino acid sequence homology (Mohammed et al., 2004). Both enzymes catalyze the first committed step in prostaglandins and thromboxanes synthesis upon cellular activation (Breyer et al., 2001). It is generally accepted that COX-1 is involved in cellular housekeeping functions necessary for normal physiological activity whereas COX-2 acts primarily at sites of inflammation, such as macrophages and monocytes. COX-2 is the form of the enzyme that has the main responsibility for the synthesis of prostanoids involved in severe inflammatory states (Stables and Gilroy, 2011). COX converts AA to PGH_2, the common precursor of all prostanoids, which is converted by prostaglandin isomerases, yielding PGE_2, PGD_2, and PGF_2, or by thromboxane synthase to create TXA_2, or by prostacyclin synthase to produce PGI_2 (Folco and Murphy, 2006).

In the LOX pathway, the major eicosanoid metabolites produced are the leukotrienes. In mammals, LOX are classified with respect to their positional specificity of AA oxygenation into 5-, 8-, 12-, and 15-LOX (Khun et al., 2002). Collectively, LOX enzymes catalyze the oxygenation of AA into hydroperoxyeicosatetraenoic acids (HPETE) and the conversion to their corresponding HETE leading to the formation of leukotrienes, lipoxins and hepoxillins (HO). 5-LOX is mainly expressed in inflammatory cells (i.e., monocytes, macrophages, and granulocytes) and catalyzes the synthesis of the unstable LTA_4, which can be converted either into LTB_4 (via the addition of a water group) or into cysteinyl-LT, LTC_4, LTD_4, and LTE_4 (via glutathione *S*-transferase) (Patterson, 2012). While the 12-LOX enzyme is found in both platelets and some epithelial cells, the 15-LOX is found only in some epithelial cells (Calder, 2001). Two different isoforms of 15-LOX have been identified in humans (His et al., 2002). The 15-LOX-1 preferentially metabolizes LA primarily to the 13-(*S*)-hydroxyoctadecadienoic (13-(*S*)-HODE) acid and also metabolizes AA to the 15-(*S*)-HETE acid, whereas 15-LOX-2 primarily converts AA to 15-HETE and hardly metabolizes LA (Chu et al., 2002). Previous observations indicate that novel eicosanoid mediators, called lipoxins, derived from the LOX pathway, are formed and play a key role in initiating the resolution of acute inflammation (Chiang et al., 2006). The discovery of lipoxins came from studies of the 15-(*S*)-HPETE metabolism, the primary metabolite of 15-LOX, in human neutrophils under the presence of calcium ionophore (Serhan et al., 1984). The ionophore stimulated the 5-LOX activation and formation of new metabolites called lipoxin A_4 and lipoxin B_4 (LXA_4 and LXB_4, respectively). These lipoxins result from the combined action of 15-LOX and 5-LOX (Serhan, 1997). The products of LOX pathway include leukotrienes, lipoxins, HO and HETE, that can act as signaling molecules or give rise to the production of reactive oxygen species (ROS).

Non-enzymatic peroxidation of membrane PUFA, during oxidative stress, may produce prostaglandin-like products called isoprostanes. F_2-isoprostanes, generated by AA metabolism, are considered to be reliable markers of oxidative stress, in particular for pulmonary disease (Lawson et al., 1999).

Although the metabolism of AA into vasoactive products by COX and LOX enzymes has been well described, a growing awareness into a less well-characterized pathway of AA metabolism, the CYP monooxygenases, has been rekindled in the last decade. It is now well-appreciated that CYP also catalyze the conversion of fatty acids, such as AA, into products including EET, HETE and DHET (Stables and Gilroy, 2011). For instance, AA is metabolized in the vascular endothelium by CYP epoxygenase to EET, which can then be converted by epoxide hydrolase to the respective regioisomer of DHET (Roman, 2002).

In the vascular smooth muscle cells, AA is catalyzed by CYP hydroxylases to 20-HETE, a vasoconstrictor metabolite (Wang et al., 1999). EET and HETE have different biological properties based on their sites of production (Capdevila et al., 2000). In contrast to leukotrienes and EET, which are generated by intracellular biosynthesis, lipoxins are generated through cell-cell interactions by a process known as transcellular biosynthesis (Serhan and Chiang, 2004; Smith, 2006). Transcellular biosynthesis would arise when donor cells are stimulated to release free AA from membrane phospholipids and, rather than immediate converted into LTA_4 or PGH_2 or reacylation back into phospholipids, the free AA is taken up by the acceptor cell followed by enzymatic oxygenation and transformation into various eicosanoid mediator pathways (Folco and Murphy, 2006). Thus, the metabolism of AA may follow multiple interrelated pathways leading to the release of a wide variety of biologically active key mediators, which are believed to play regulatory roles in the normal differentiation process, as well as, in the pathogenesis of specific diseases.

2.1. Biological Functions

AA has multiple physiological roles in living cells. It is an important fatty acid for the structural and functional integrity of cellular membranes, such as thickness and fluidity, activity of membrane associated proteins (enzymes, ion channels, receptors and transporters) and, thereby, overall cell function. AA influences the activity of second messengers and is also the endogenous precursor of a multitude of vasoactive eicosanoids affecting signal transduction and gene transcription, cell proliferation and apoptosis (Keane et al., 2011; Kremmyda et al., 2011).

The selective actions of free AA may be explained simply by its specific release under physiological conditions or by the absence of such mechanisms for releasing other LC-PUFA, compounds that might otherwise share its biochemical effects (Brash, 2001). The well-recognized functional role of AA is as a cell signaling molecule, either in its original form or after its conversion to eicosanoids that are important to maintain tissue homeostasis (Haeggström et al., 2010). Although the physiological effects of AA are often attributable to its conversion by cellular oxygenases (COX, LOX and CYP) to prostaglandins, leukotrienes and other bioactive metabolites, unconverted AA seems to possess its own biological activity (Brash, 2001). To distinguish the effects of AA from those of its derivatives, the non-metabolizable AA analogue 5,8,11,14-eicosatetraynoic acid (ETYA) is often used. Earlier studies showed proliferative effects of AA and ETYA on bovine aortic endothelial cells, most likely via the induction of store-independent calcium influx, which has a mitogenic action (Fiorio Pla and Munaron, 2001).

AA is metabolized so that it may be used to modulate ion channel activities, protein kinases with subsequent phosphorylation of target molecules, and neurotransmitter uptake systems. Many of AA target molecules are involved in neuronal development. AA is one of the most abundant fatty acid in the brain, and is present in similar quantities to DHA. AA helps to maintain hippocampal cell membrane fluidity (Fukaya et al., 2007). It also helps to protect the brain from oxidative stress by activating peroxisome proliferator-activated receptors (PPAR), which possess transcriptional activity (Wang et al., 2006). AA also activates syntaxin-3, a protein involved in the growth and repair of neurons (Darios and Davletov, 2006). Earlier reports have also stressed the physiological role of AA in the cardiovascular system, namely, in human platelets as it is a precursor of TXA_2, which induces platelet aggregation (Moncada and Vane, 1979; Jubelirer et al., 1980). Further evidence of the actions of AA and other PUFA includes the induction of apoptosis (Artwohl et al., 2004; Fang et al., 2008). In human endothelial cells, saturated stearic acid concentration dependently induced apoptosis, whereas LA, ALA, and AA triggered programmed cell death only at a highest concentration (Artwohl et al., 2004). A potential mechanism relates to the activation of sphingomyelin hydrolysis by free AA and the subsequent induction of apoptosis by ceramide (Jayadev et al., 1994).

AA is a key inflammatory intermediate through several mechanisms, including cell surface and intracellular receptors/sensors that control inflammatory cell signaling and gene expression patterns (Calder, 2012). Important regulators of metabolic and inflammatory signaling, in both metabolic and immune disorders, are the PPAR family members (Bensinger and Tontonoz, 2008). In the regulation of inflammatory signaling, PPAR inhibits sterol regulatory element-binding protein (SREBP)-1c and nuclear factor-κB (NF-κB) expression (Patterson et al., 2012). However, the *n*-3 PUFA are more potent inhibitors of SREBP-1c than the *n*-6 PUFA (Schmitz and Ecker, 2008). NF-κB, another transcription factor regulated by PUFA, is found in animals in almost cell types, and regulates the immune response to infection. It controls several cytokines (e.g., interleukin (IL)-1, and tumor necrosis factor (TNF)-α), chemokines (e.g., IL-8, monocyte chemoattractant protein-1), adhesion molecules, and inducible effector enzymes (e.g., inducible nitric oxide (NO) synthase and COX-2) (Wall et al., 2010). Summing up, greater amounts of AA in cell membrane phospholipids result in increased COX and LOX enzyme production of AA-derived eicosanoids and decreased EPA/DHA-derived eicosanoids, leading to pro-inflammatory cytokine production and overall inflammation. This, in turn, decreases PPAR-α mRNA levels, while increasing both SREBP-1c and NF-κB gene expression. This change in gene expression promotes inflammation in several metabolic disorders and immune related diseases (Patterson et al., 2012).

2.2. Eicosanoids and Their Role on Inflammation

Eicosanoids (preferred IUPAC name icosanoids) belong to a wide range and complex family of bioactive lipid mediators, primarily derived from AA oxidative pathway but also from EPA and DHA pathways. These compouds regulate a multiplicity of cellular functions, especially those concerning inflammation and immunity (Tilley et al., 2001; Harris et al., 2002; Harizi et al., 2008). Eicosanoids exert their effects mainly locally through interaction with specific receptors on the cell surface or nuclear membrane and act at nanomolar concentrations in an autocrine/paracrine manner on target cells to elicit a diverse set of pharmacological effects (Flachs et al., 2009). Cells of the innate immune system are major producers of eicosanoids. While prostaglandins are produced by practically all cells of the body, leukotrienes are synthesized predominantly by inflammatory cells such as polymorphonuclear leukocytes, monocytes, macrophages, mast cells and dendritic cells. Indeed, these lipid mediators are involved in modulating the intensity and duration of the inflammatory response, have cell- and stimulus-specific sources and frequently opposing effects (Tilley et al., 2001; Calder, 2006). Inflammation, which can be acute or chronic in nature, is a normal host defense mechanism that protects from infection and helps to repair the damaged site and to restore tissue homeostasis (Bederska-Lojewska et al., 2013). It begins when pro-inflammatory substances such as cytokines, thrombin, or ROS are released in response to various tissue disorders (Bogatcheva et al., 2005). Because inflammatory cells typically contain a high proportion of AA compared to other 20 carbon PUFA (DGLA and EPA), AA is usually the main substrate for eicosanoid synthesis (Calder, 2008; Calder, 2009). How AA is metabolized in the body dictates its pro- or anti-inflammatory function. As shown in the Table 1, some AA-derived eicosanoids have both pro- and anti-inflammatory roles, while others regulate the development and resolution of inflammation in the immune response (Ricciotti and Fitzgerald, 2011). Among the classical eicosanoids, prostaglandins were the first to be discovered, and might act as both pro- and anti-inflammatory mediators depending on the context. This is partially due to the array of prostaglandins receptors with different signal transduction pathways (Harizi et al., 2008).

In response to injury or stress, EPA-derived prostaglandins and leukotrienes are able to modulate the pro-inflammatory effects of AA-derived 2-series prostaglandins and 4-series leukotrienes, usually in the same target cells. Although similar in structure, the 2-series of prostaglandins is considered to be more mitogenic and pro-inflammatory compared with the 3-series prostaglandins derived from EPA (Bagga et al., 2003; Moreno, 2009). COX-2-derived prostaglandins promote inflammation and pain perception by increasing vascular permeability and vasodilatation, and by directing the synthesis and migration of pro-inflammatory cytokines, such as IL-1, IL-6, IL-8, TNF-α and adhesion molecules, into the site of inflammation (Tilley et al., 2001). Thromboxanes, produced in platelets, constrict blood vessels and promote platelet aggregation, an early step in blood clotting.

HPETE and leukotrienes are also a key group of pro-inflammatory mediators synthesized from LOX catalysed AA metabolism. Leukotrienes have been identified in a variety of inflammatory and allergic reactions like rheumatoid arthritis, inflammatory bowel disease, psoriasis, allergic rhinitis and bronchial asthma (Werz and Steinhilber, 2005).

Table 1. The best known pro- and anti-inflammatory mediators of AA, EPA and DHA

n-6 PUFA				
	Pro-inflammatory		*Anti-inflammatory*	
AA	Prostaglandin D_2	(Folco and Murphy, 2006)	Prostaglandin E_2	(Miles et al., 2002; Vachier et al., 2002)
	Prostaglandin E_2	(Chang and Chang, 2003; Flachs et al., 2009)	Lipoxin A_4	(Nigam et al., 1990; Godson et al., 2000)
	Prostaglandin F_2	(Folco and Murphy, 2006)	Lipoxin B_4	(Nigam et al., 1990; Godson et al., 2000)
	Thromboxane A_2	(Folco and Murphy, 2006; Flachs et al., 2009)		
	Prostacyclin I_2	(Folco and Murphy, 2006; Flachs et al., 2009)		
	Leukotriene B_4	(Chang and Chang, 2003; Flachs et al., 2009; Moreno, 2009; Thusgaard et al., 2009; Hagi et al., 2010)		
n-3 PUFA				
	Pro-inflammatory		*Anti-inflammatory*	
EPA	Not known		Prostaglandin E_3	(Bagga et al., 2003; Bachi et al., 2009)
			Leukotriene B_5	(Bachi et al., 2009; Moreno, 2009; Thusgaard et al., 2009; Hagi et al., 2010)
			Resolvin E_1	(Arita et al., 2005; Tjonahen et al., 2006; Herrera et al., 2008)
			Resolvin E_2	(Tjonahen et al., 2006)
DHA	Not known		Resolvin D_1	(Ariel and Serhan, 2007; Claria et al., 2012)
			Resolvin D_2	(Claria et al., 2012)
			Resolvin D_3	(Ariel and Serhan, 2007; Dalli et al., 2013)
			Resolvin D_4	(Ariel and Serhan, 2007)
			Protectin D_1	(Levy et al., 2007)

(adapted from Kelly et al., 2013).

In general, AA-derived prostanoids and leukotrienes have been, by definition, described as pro-inflammatory eicosanoids (Harizi et al., 2008). However, this turns out not to be entirely true (Calder, 2009). One example is the paradoxal case of PGE_2. Firstly, PGE_2 is a potent inhibitor of the production of two classic pro-inflammatory cytokines, TNF-α and IL-1 (Miles et al., 2002). Secondly, in the last few years, it was discovered that PGE_2 inhibits 5-LOX, decreasing the 4-series leukotrienes, and induces 15-LOX, promoting the formation of lipoxins (Vachier et al., 2002) with anti-inflammatory and proresolution properties (Gewirtz et al., 2002; Serhan et al., 2003). A mechanism has been proposed whereby a coordinated program for resolution initiates in the first few hours after the inflammatory response. A switch occurs whereupon the AA-derived prostanoids, which have set the inflammatory to begin, undergo further metabolism to become another generation of eicosanoids derived from AA, the lipoxins, and hence terminate inflammation at the local contained sites (Serhan and Savill, 2005). Since lipoxins are involved in the resolution of the acute inflammation that occurs as a result of the overproduction of the pro-inflammatory eicosanoids derived from AA, they are said to have proresolving and anti-inflammatory functions. Lipoxins properties are similar to the less inflammatory actions of *n*-3 PUFA-derived 3-series prostaglandins and 5-series leukotrienes, and to a novel family of potent anti-inflammatory lipid mediators called resolvins, maresins and protectins (Calder, 2006; Tjonahen et al., 2006). This new family has been termed proresolving mediators since they are involved in the clearance and regulation of inflammatory substances (Bannenberg and Serhan, 2010).

Accordingly to the above considerations, the manipulation of AA metabolism represents a promising approach to develop chemopreventive strategies in a variety of health conditions. Since eicosanoids are mediators of several specific diseases such as cancer, cardiovascular disease and atherosclerosis, or asthma, many drugs that disturb eicosanoid synthesis have been developed and are widely used. In fact, AA metabolism is a long recognized target for anti-inflammatory agents and the efficacy of non-steroidal anti-inflammatory drugs (NSAID), such as aspirin and ibuprofen, which are well accepted inhibitors of COX activity, indicates the importance of this pathway in the pathophysiology of inflammation associated diseases (Serhan and Savill, 2005; Calder, 2009). If indeed aspirin blocks AA metabolism by COX-2, other PUFA metabolites can bind, thus resulting in the potent anti-inflammatory/pro-resolving aspirin-triggered resolvins, EPA and DHA-derived (Serhan and Savill, 2005). Now, the question is: does the effectiveness of aspirin depend on the amount of AA present, that is, does aspirin work best when the ratio of *n*-6/*n*-3 is high? This remains to be elucidated.

3. Dietary Sources and Nutritional Aspects

Dietary imbalance in the consumption of *n*-6 and *n*-3 PUFA is representative of the Western diet. LA, the most prevalent *n*-6 fatty acid in the human diet, is considered a major source of AA. Thus, AA is obtained directly from dietary sources or synthesized in the body from LA. Excessive intake of LA inhibits the production of EPA and DHA, and leads to the overproduction of AA. Within the last decades, changing nutritional habits in Western communities led to a fourfold increase in the supply of dietary AA, caused by the same increase in the consumption of meat and meat products (Adam, 1992). Thus, foods from animal origin, including meat, poultry and eggs, are the main dietary sources of AA. The level of AA in lean beef and lamb is 30-40 mg/100 g, being 1.5-2.5 times lower compared to its content in chicken, duck, turkey and, also, pig (43, 99, 74 and 54 mg/100 g, respectively) (Li et al., 1998). Based on data from the 2005-2006 National Health and Nutrition Examination Survey (NHANES), chicken and beef contribute with 26.9 and 7.3% of the total AA intake, respectively. Evidences suggest a daily intake of 50-300 mg of AA for adults in Western societies (Sinclair and O'Dea, 1993; Jonnalagadda et al., 1995; Mann et al., 1995; Nelson et al., 1997a; Sioen et al., 2006; Elmadfa and Kornsteiner 2009; Kawabata et al., 2011) with some reports estimating a daily level up to 500 mg (Taber et al., 1998). A scientific advisory from the American Heart Association has favorably evaluated the health impact of dietary *n*-6 fatty acids, including AA. The committee does not recommend limiting this fatty acid intake. In fact, the American Heart Association supports that individuals should follow a diet containing at least 5-10% of energy coming from *n*-6 PUFA, among other dietary recommendations (Harris et al., 2009). AA supplementation of healthy adult's diets appears to offer neither toxicity nor significant safety risk. Daily dosages around 1.5 g of AA have been well tolerated during several clinical studies, with no significant side effects on immune function, serum lipids, platelet aggregation, kidney and liver function (Kelley et al., 1997; Nelson et al., 1997abc;Kelley et al., 1998; Thies et al., 2001ab; Kusumoto et al., 2007). Furthermore, high concentrations of AA in muscle tissue were found correlated with improved insulin sensitivity (Borkman et al., 1993). As reported by Calder (2007), taken together, these studies suggest that, rather than being harmful, moderately increased AA intake is probably harmless in healthy adults, although the effect of intakes above 1.5 g/d are not well known, as well as, the increased intake in diseased individuals.

Dietary AA by itself increases the AA content in plasma in a dose-dependent manner, most probably due to a low oxidation rate and a high affinity uptake system (Adam, 1992). Yet, this linear response is not verified for the conversion of dietary LA into plasma AA (Mohrhauer and Holman, 1963). AA is preferentially incorporated into phospholipids (8%) of plasma membranes, and less into triacylglycerols (1.64%), where it is found exclusively in the *sn*-2 or -3 positions (Breckenridge et al., 1976; Spector, 2000). This positional specificity results in a resistance of AA to some enzymes' action, namely lipoprotein lipase and hepatic lipase, which cleave fatty acids from the *sn*-1 position for oxidation (Morley and Kuksis, 1972). An early report describes the time course of AA in plasma lipids during and after diet supplementation with 6 g of ethylarachidonate/d for 3 weeks (Seyberth et al., 1975). A dose-response increase in AA percentage was observed in all lipid classes (triacylglycerols, phospholipids, and cholesterol esters), except for free fatty acids. During the observation times, no saturation of the AA pool and transport systems, and hence no steady state, were reached.

AA is also present in human breast milk and plays a central role for maternal, fetal and infant health (Birch et al., 2000; Clandinin et al., 2005). The fatty acid composition of placenta is indicative of maternal fatty acid status and determines fatty acids' transfer to the fetus. The placenta is mainly composed of phospholipids, which comprise about 88% of total lipids, dominated by PUFA, followed by saturated fatty acids (mainly palmitic and stearic acids), and monounsaturated fatty acids (mainly oleic acid) in lower percentages. Noticeably, AA is the major PUFA in placenta tissue. Its levels (about 22-25%) are higher than that of DHA and EPA, as well as LA and DGLA (Kremmyda et al., 2011). Maternal breast milk is of great importance in humans and other mammals since it is the main source of all nutrients, including lipids for the newborn. Breast milk fatty acids content is important for brain development, cognition, vision, as well as, immune development of the infant. It is very much dependent on maternal dietary fat intake during gestation which is deposited in the adipose tissue and released into milk during lactation (Dunstan et al., 2007). Human milk provides mainly saturated (palmitic acid) and monounsaturated (oleic acid) fatty acids (Jensen, 1999), while DHA (0.2-1%) and AA (0.3-0.7%) are the dominant LC-PUFA, with DHA being more variable and sensitive to maternal dietary intake (Brenna et al., 2007). AA increases in breast milk following oral consumption, either via diet or supplementation, in a relatively dose-dependent manner (Weseler et al., 2008) although supplemental DHA (fish oil) may reduce the concentration of AA in breast milk (van Goor et al., 2009). According to the former authors, this increase has been quantified as high as 14-23% over 2-12 weeks (220 mg of AA supplementation) while one week of 300 mg of AA has been proven insufficient to increase concentrations significantly (Smit et al., 2000). van Goor and colleagues (2009) concluded that dietary AA intake becomes noticeable after prolonged supplementation.

AA also appears to be an important constituent of infant formulas, being required for optimal growth and development (Makrides et al., 1995; Koletzko et al., 1996). The supplementation of LC-PUFA to infant formulas has been shown to enable preterm infants to achieve immune development similar to that seen with breast-milk feeding (Field et al., 2000). AA-enriched triacylglycerol oil (AA-oil) derived from the fungus *Mortierella alpine* has been widely used as a food additive for infant formulas. The AA-oil is not mutagenic in chromosomal aberration assays performed with Chinese hamster ovary cells (Arterburn et al., 2000a). Furthermore, neither maternal toxicity nor teratogenic potential were seen in rats administered up to 2.5 g/kg body weight/d of this oil (Arterburn et al., 2000b). Likewise, no toxicity was observed in rats fed on repeated doses of AA-oil for short, 28 days, and long, 90 days, periods of treatment (Hempenius et al., 1997; Hempenius et al., 2000). Moreover, when AA levels in the brain, heart, and liver were increased by 8%, 59% and 76%, respectively, as a result of AA-oil gavage (2.5 g/kg/d), no developmental, histopathological nor neuropathological differences were seen between rats fed on AA versus vehicle controls (Koskelo et al., 1997).

AA is also commercially available as a dietary supplement for athletes to enhance training performance. This is due to its role in growth and repair of skeletal muscle tissue (Trappe et al., 2001). It is a bioactive compound involved in myogenic inflammation that occurs in response to mechanical strain, such as resistance-training (Vandenburgh et al., 1993). Indeed, AA is the metabolic precursor of prostaglandins, which can *regulate* muscle cell growth by influencing multiple steps of myogenesis (Horsley and Pavlath, 2003). Markworth and Cameron-Smith (2013) investigated the direct effect of free AA supplementation and subsequent eicosanoid biosynthesis on skeletal myocyte growth *in vitro*. The former authors verified that AA-stimulated growth was associated with markedly increased secretion of PGE_2 and $PGF_{2\alpha}$. However, incubation of myocytes with prostaglandins-rich conditioned medium failed to mimic the effects of direct AA supplementation. Previous studies examining the inhibition of the COX-2 pathway have discerned that

COX-2 isozyme products mitigate protein degradation and subsequent myofibrillar regeneration in the skeletal muscle (Goldberg et al., 1984; Bondesen et al., 2004). Moreover, the first clinical study on the use of AA as a sport supplement found that AA enhances anaerobic capacity, although it was not able to improve strength, muscle mass, hormones of anabolism, or intracellular markers of muscle hypertrophy (Roberts et al., 2007). Nevertheless, preliminary data were found to support AA as a nutritional ergogenic (Roberts et al., 2007).

4. Implications of Arachidonic Acid in Human Health

4.1. Cancer

Cancer remains a major health problem despite many scientific advances towards therapeutic treatment. Dietary habits are well recognized as an important contributor to cancer risk, and increased fat consumption, in particular, has been associated with the development of specific types of cancer, such as colorectal, breast, prostate and skin (Azrad et al., 2013). As inflammation is a key factor of tumorigenesis, anti- and pro-inflammatory mechanisms mediated by PUFA have been identified on cancer development. The lipid mediators derived from AA most linked with cancer are the prostaglandins and leukotrienes as they have been shown to play important roles in the progression of cancer through angiogenesis, cell proliferation, metastasis and apoptosis (Wang and DuBois, 2010). In cancer cells, COX-2 is frequently overexpressed resulting in the production of high levels of PGE_2 (Wang and DuBois, 2010). The NSAID and selective COX-2 inhibitors may exert chemoprotective actions by targeting the COX pathway and decreasing the synthesis of PGE_2 (Mahmud et al., 2004; Takkouche et al., 2008). Under normal physiological conditions, the 5-LOX enzyme is not expressed but, in turn, during inflammation and tumorigenesis it converts AA to LTB_4 (Cathcart et al., 2011). By using specific inhibitors of different pathways, Ghosh and Myers (1997) determined that the metabolic conversion of AA through 5-LOX is required for its growth stimulatory effect on human prostate cancer cells. Besides these biological findings about the relationship between AA intake, its level in blood or tissues, and cancer development, it is unclear whether dietary AA intake leads to cancer in humans.

In fact, it has been suggested that some AA mediators inhibit tumor growth (Mahmud et al., 2004; Chiang et al., 2006; Takkouche et al., 2008). Apart from the COX pathway, there are AA metabolites that may decrease cancer risk. LXA_4 is considered an anti-inflammatory mediator produced by the LOX pathway and based on its anti-angiogenic properties, is regarded as a tumor growth suppressor (Chiang et al., 2006; Chen et al., 2011). In H22 cells, LXA_4 inhibited the production of vascular endothelial growth factor and reduced hypoxia-inducible factor-1α level (Chen et al., 2011). Moreover, aspirin or NSAID may not only inhibit the COX metabolites production, but also redirect AA into the LOX pathway (Sakai et al., 2012). Other results suggested that the protective effect of NSAID on the incidence of colon carcinoma was attributable to AA accumulation, rather than to a reduction on prostaglandins levels (Chan et al., 1998). AA and its metabolites have been shown to inhibit the growth of human breast cancer cell lines, through PPAR-α involvement (Bocca et al., 2008). Alternativelly and considering apoptosis as a major player in tumorogenesis, AA and EPA blocked proliferation of the promyelocytic leukemic HL-60 cells (Brash, 2001).

Some animal studies revealed that AA administration did not affect tumor growth (Petrik et al., 2000; Imai et al., 2012). Petrik et al., (2000) observed that AA had no effect on tumor load in ApcMin/1 mice despite increasing tissue AA concentration and *ex vivo* prostaglandins production. In a medium-term multi-organ carcinogenesis assay in rats, AA had no tumor promoting activity in any organs or tissues initiated with five carcinogenic compounds (Imai et al., 2012).

Among human studies, controversial observations about AA exposure and colorectal cancer risk are found. Lin et al., (2004) suggested no association between them. Nkondjock et al., (2003a) and Murff et al., (2009) reported a positive dose-response relationship between dietary AA and colorectal cancer risk, while other studies observed the inverse (Kuriki et al., 2006; Ghadimi et al., 2008). To illustrate this, a case-control study in Japan demonstrated that serum AA was able to reduce up to 59% the colorectal adenoma cancer risk (Ghadimi et al., 2008).

For prostate cancer, a clear positive association between increased cancer risk and AA exposure was not found (Männistö et al., 2003; Chavarro et al., 2007; Crowe et al., 2008). Cross-sectional studies in USA, Nigeria and Greece, reported significant decreases of tissue AA levels in prostate cancer subjects (Mamalakis et al., 2002; Faas et al., 2003; Ukoli et al., 2010).

While epidemiological studies suggest that dietary *n*-3 PUFA may exert anti-tumor effects on breast cancer progression, the reports relative to *n*-6

PUFA anti-carcinogen effects remain unclear. Observational studies that investigated the relation of dietary AA intake and breast cancer risk did not find significant changes between them (Voorrips et al., 2002; Nkondjock et al., 2003b). Moreover, blood and tissue AA levels were not associated with higher breast cancer risk (Aro et al., 2000; Maillard et al., 2002; Saadatian-Elahi et al., 2002).

In what skin cancer concerns, it is also difficult to draw definitive conclusions. Harris et al., (2005) observed that increased levels of AA in red blood cell membranes were associated with increased risk of skin carcinoma. However, Hakin et al., (2000) observed a consistent pattern for marginally lower risk of skin squamous cell carcinoma with increasing intakes of AA although the trends were not statistically significant.

The anti-inflammatory properties of some AA mediators may help to explain the inconsistent results correlating AA with cancer risk. Two other aspects may contribute for this poor evidence. Firstly, blood and tissue levels of AA may not always represent the AA intake, because, although not consensual, some authors verified a low correlation between dietary and AA contents in adipose tissue and serum phospholipids (Garland et al., 1998; Kobayashi et al., 2003). Secondly, AA metabolites may not be proportional to blood or tissue AA levels (Kakutani et al., 2011).

4.2. Cardiovascular Diseases and Atherosclerosis

Inflammation plays an important role in coronary heart disease (CHD) as participates in atherosclerotic events. Under normal physiological conditions, endothelial cells synthesize adequate amounts of NO, PGI_2, and PGE_1, preserving a balance between pro- and anti-inflammatory molecules. In atherosclerosis, a hyperproliferative disease associated with dysregulation in lipid metabolism and vascular inflammation, this balance is disturbed. The pro-inflammatory cytokines, such as IL-1, IL-2, IL-6 and TNF-α, are increased and can induce oxidative stress by enhancing the production of ROS by monocytes, macrophages, and leukocytes, contributing to the further progression of the disease (Libby, 2006). Since PUFA and their eicosanoid derivatives modulate inflammation, they naturally play a significant role in this disease. AA is precursor of pro-inflammatory, vasoconstrictive and pro-aggregatory molecules, such as PGE_2, TXA_2, and LTB_4. PGE_2, among several effects, plays a pivotal role in cardiovascular inflammation (Furuyashiki and Narumiya, 2011). TXA_2 acts as a potent vasoconstrictor and a powerful activator of platelet aggregation (Moncada and Vane, 1979; Jubelirer et al., 1980; Sellers and Stallone, 2008). LTB_4 induces a complex cascade of molecular and cellular events including the promotion of leukocyte adherence to the vascular endothelium and generation of ROS (Aiello et al., 2002). In addition, LOX are lipid-peroxidizing enzymes that have been implicated in atherogenesis. Among the six functional human LOX-isoforms, four, the 12, the 15, the platelet-type 12, and the 5, have been implicated in atherogenesis (Kuhn et al., 2007).

In theory, reduced *n*-6 intake, with subsequent reductions in tissue AA content, would reduce the pro-inflammatory state and, therefore, lower the risk for CHD. However, AA also produces anti-inflammatory and anti-aggregatory metabolites, such as LXA_4, prostacyclin and EET acids (Node et al., 1999; Serhan, 2005). In particular, EET acids have important vasodilator properties through hyperpolarization and relaxation of vascular smooth muscle cells (Oltman et al., 1998). A large number of human studies suggest that higher intake of AA reduces the risk for CHD. The American Heart Association has been focused on the association between *n*-6 PUFA, including AA, and the risk of CHD (Harris et al., 2009). Data collected from randomized trials, case-control and cohort studies, and long-term animal feeding experiments indicate that the consumption of at least 5% of energy from *n*-6 PUFA reduces the risk of CHD relative to lower intakes (Harris et al., 2009). In an epidemiological study, high plasma levels of AA were associated with decreased levels of the pro-inflammatory markers IL-6 and IL-1 receptor antagonist, and increased levels of the anti-inflammatory marker transforming growth factor-β (TGF-β) (Ferrucci et al., 2006). High-AA diets (1.7 g/d) supplemented to healthy male volunteers did not alter serum lipid levels and produced no physiological change in blood coagulation and thrombotic parameters (Nelson et al., 1997bc). Likewise, supplementation with AA had no adverse effects on human immune response and platelet function (Kelley et al., 1997; Kusumoto et al., 2007). Interestingly, AA seems to improve mitochondrial function in rat´s heart (Khairallah et al., 2012).

On the other hand, aging is a recognized contributing factor for the development of cardiovascular disease due to endothelial dysfunction characterized by progressive decline in endothelium-dependent vasorelaxation of arteries (Matz et al., 2000). Little is known on the involvement of membrane's AA content in age-related cardiovascular dysfunction. Japanese elderly individuals were tested daily with 240 mg of AA for 3 months (Oe et al., 2008). In periods of increased blood flow, but not at rest, AA was able to improve coronary flow velocity (Oe et al., 2008). In Fisher rats, AA appears to completely reverse the age-related increase in vasoconstriction induced by phenylephrine (Nakano et al., 2007). In the same study, AA was able to slightly augment an acetylcholine-induced vasorelaxation effect in aged rats with no apparent benefit to young rats (Nakano et al., 2007).

4.3. Brain and Neurodegenerative Diseases

AA is a predominant PUFA in all types of cells throughout the different regions of the brain. The release of AA from membrane phospholipids can serve as an intercellular messenger in the central nervous system, which may regulate vital aspects of neuronal membrane biology, including protein-lipid interactions, synaptic signaling (Kaufmann et al., 1996; Kawasaki et al., 2002; Rapoort, 2008), neurotransmitter release (Ojeda et al., 1989), neuronal gene expression (Lerea et al., 1997), cerebral blood flow (Pickard, 1981) and sleep-wake regulation (Hayaishi and Matsumura, 1995). AA and prostaglandins regulate long-term potentiation which is important for learning and memory (Williams et al., 1989). AA supplementation seems to be effective in preserving the cognitive ability by conserving the nerve synapses plasticity in aged animals (McGahon et al., 1997; Kotani et al., 2003). In one study funded by the U.S. National Institute of Child Health and Human Development, infants (18 months old) given supplemental AA during 17 weeks demonstrated significant improvements in intelligence (Birch et al., 2007). Giving 240 mg/d of AA to elderly men produced marked improvements in cognitive dysfunction (Ishikura et al., 2009). Another study reported improvements in mild cognitive impairment in elderly patients taking 240 mg of AA combined with 240 mg of DHA, particularly with regard to memory and attention scores (Kotani et al., 2006).

In contrast, a series of highly reactive prostaglandins, eicosanoids, free fatty acids, lysophospolipids and ROS, all generated by enhanced PLA_2 activity and AA release, participate in cellular injury, particularly in neurodegeneration (Bazan et al., 2002). Alterations in brain AA metabolism have been implicated in a number of neurological, neurodegenerative and psychiatric disorders.

Epilepsy and ischemia are highly affected by AA end-products, namely COX-2 (McCown et al., 1997; Sapirstein and Bonventre, 2000). PLA_2 is an important mediator of ischemic and oxidative injuries in the brain. In fact, cytosolic PLA_2 deficient mice are protected against neuronal injury when compared to their wild type controls (Sapirstein and Bonventre, 2000).

Inhibitors of both COX and LOX pathways have been shown to reduce cerebral deficits following ischemia and trauma (Phillis and O'Regan, 2003).

Excessive levels of AA and prostaglandins can contribute to excitotoxicity, which may be involved in Alzheimer's pathogenesis (Phillis et al., 2006). First described a century ago, Alzheimer's disease prevalence has markedly increased, and no effective treatment exists. One of the hallmarks of the Alzheimer's disease is the accumulation of protein fragments between neurons (Masters et al., 1985). In healthy brain, β-amyloid protein fragments are broken down and eliminated. In Alzheimer's disease brain, these fragments accumulate to form insoluble plaques and additional chronic inflammation has been observed (Townsend and Praticò, 2005). AA may contribute to Alzheimer's disease inflammation through leukotrienes which are potent chemoattractants for monocytes and granulocytes. In a mouse model of Alzheimer's disease, there was an increase in AA and its metabolites in brain tissue and it was observed that PLA_2 reduction ameliorates cognitive deficits (Sanchez-Mejia et al., 2008). In fact, NSAID can exert neuroprotective effects in Alzheimer's disease by suppressing AA metabolic pathway (Christie et al., 1999). Recently, the work of Amtul and colleagues (2012) supports the involvement of AA and its metabolites in the production of β-amyloid peptides, providing for the first time, the structural argument for their detrimental effects on Alzheimer's disease biochemical basis. In contrast to these studies, the supplementation of AA during the early stages seems to be effective in reducing symptoms and slowing the progression of the disease (Schaeffer et al., 2009).

The Parkinson's disease is also a progressive neurodegenerative disorder characterized by slow movements, tremor, and walking impairment due to loss of midbrain nigrostriatal neurons and depletion of striatal dopamine (Jakowec and Petzinger, 2004). The COX-2 inhibition prevents the formation of the oxidant complex dopamine-quinone, which has been implicated in the pathogenesis of this disease (Teismann et al., 2003). A case-control study examined the relationship between dietary intake of individual fatty acids and the risk of Parkinson's disease in Japan (Miyake et al., 2010). It was concluded that higher AA intake may be related to an increased risk of this disease (Miyake et al., 2010). COX-2 upregulation is also essential in the neurodegenerative process involved in amyotrophic lateral sclerosis (Almer et al., 2001).

Clinical research suggests that AA cascade may be determinant in mood disorders (Bazinet, 2009). Since AA suppresses neuronal plasticity, it is hypothesized that the neuroprotective effects of lithium and valproate are achievable through downregulation of AA pathway (Sublette et al., 2004). *In vivo*, PGE_2 has been shown to be elevated in depression, and to have profound effects in sleep disorders, appetite, fatigue, social interactiveness, psychomotor retardation, and poor concentration. The celecoxib, a selective COX-2 inhibitor drug, may produce a rapid-onset mood improvement in bipolar patients and also displays beneficial effects in schizophrenia disorder (Muller et al., 2002; Nery et al., 2008).

4.4. Obesity

The incidence of obesity and its related metabolic disorders has increased significantly over the past three decades, culminating in the current global epidemic and leading to the search for contributing factors. Exposure to a typical Western diet increases the risk of obesity and metabolic disorders throughout the life-course, creating an intergenerational cycle of metabolic disease. This epidemic of metabolic disease has coincided with a marked increase in the intake of *n*-6 polyunsaturated fatty acids, leading to suggestions that the two may be causally related. The pro-adipogenic properties of *n*-6 PUFA in parallel with low intake of *n*-3 PUFA result in a progressive accumulation of body fat across generations (Muhlhausler and Aihaud, 2013) and low-grade chronic inflammation (Simopoulos, 2006) predisposing humans to obesity (Cordain et al., 2005; Simopoulos, 2006).

Low-grade chronic inflammation, like the one observed in the obese state, is persistent, unresolved and characterized by macrophage infiltration perhaps due to additional adipocyte cell debris (Cancello and Clement, 2006; Cinti, 2011). On the other hand, uninterrupted adipogenic differentiation may lead to adipose upgrowth. One such study (Laschober et al., 2011) suggests that age-associated inflammation or low-grade chronic inflammation may increase adipogenesis at the expense of other cell types; whether this is the result of dietary habits and other lifestyles in industrialized countries or not, an evolutionary reason may exist to explain why fat cell production increases over time (Kelly et al., 2013). In the field of nutritional lipid research, it is currently hipothesized that a poor dietary *n*-3 PUFA intake, resulting in constant and augmented production of *n*-6 PUFA-derived metabolites (Folco and Murphy, 2006; Kelly et al., 2013) might be an additional cause for this metabolic deregulation. To illustrate this sentence, it was shown that a higher plasma concentration of *n*-3 PUFA accelerates wound healing in non-obese healthy subjects (McDaniel et al., 2008). Because lipoxins are involved in wound healing, by recruiting monocytes without raising inflammation (Stables and Gilroy, 2011), it may indicate that *n*-3 PUFA work synergistically with the *n*-6 PUFA-derived lipoxins.

As being so, evidence from animal and human studies supports the premise that lowering the ratio of *n*-6/*n*-3 PUFA by increasing *n*-3 PUFA decreases adipocyte differentiation (Kim et al., 2006) and benefits body fat composition (Flachs et al., 2005) through anti-inflammatory mechanisms. PGE_2 and LTB_4 are the key *n*-6 PUFA metabolites in the inflammatory response, and both are strongly associated with obesity (Flachs et al., 2009). Given that the Western diet is believed to be inflammatory, dietary *n*-3 PUFA guidelines should take into account their fundamental anti-inflammatory properties, in reducing systemic and local adipose tissue inflammation as well as macrophage infiltration in obese models (Kelly et al., 2013). 5-LOX oxidizes AA and EPA to produce the pro-inflammatory 4-series (LTB_4) (Werz and Steinhilber, 2005) and the anti-inflammatory 5-series (LTB_5), respectively (Kobayashi et al., 1995). The supplementation of overweight subjects with 640 mg of EPA and 480 mg of DHA for 6 weeks resulted in a significant reduction in neutrophil LTB_4 production, while there was a significant increase in LTB_5 production in the treatment group compared to the control (Nielsen et al., 2012). LTB_5 can inhibit LTB_4 production, showing that it is anti-inflammatory in nature. Hereupon, 5-LOX is becoming a target for anti-obesity drugs because it is significantly elevated in both ob/ob mice (López-Parra et al., 2008) and obese humans (Kaaman et al., 2006). Furthermore, 5-LOX activating protein mRNA expression decreases after weight loss (Kaaman et al., 2006) and 5-LOX is strongly associated with obesity-associated inflammation (Martinez-Clemente et al., 2011).

Although it is clear that there is a shift in basal inflammatory levels over time, coinciding with an increase in fat mass, evidence is still being gathered to provide the link between body fat and low-grade chronic inflammation. The hypothesis is that a high intake of *n*-6 PUFA coupled with a low intake of *n*-3 PUFA promotes low-grade chronic inflammation (Kelly et al., 2013), thus contributing to obesity. Therefore, the ultimate goal in the forthcoming years would be to elucidate if reducing, or flattening out, the inflammation, using *n*-3 PUFA intake, would inhibit accrual of fat mass over time. Moreover, this may answer at what age *n*-3 PUFA intervention would be most beneficial to fat upgrowth.

4.5. Nonalcoholic Fatty Liver Disease

Nonalcoholic fatty liver disease (NAFLD) is often described as the hepatic component of the metabolic syndrome and is rapidly becoming a serious public health problem (Byrne et al., 2009). The range of liver damage associated with NAFLD begins with steatosis and can often persist to further steatohepatitis (NASH), advanced fibrosis and cirrhosis (Byrne, 2010). Moreover, NAFLD is an independent risk factor for cardiovascular disease. The occurrence of NAFLD is much higher in subgroups of the population with obesity, metabolic syndrome and type 2 diabetes, whereby the prevalence in developing the disease for those with type 2 diabetes may be as high as 70% (Erickson, 2009; Byrne, 2010). Both nutritional factors and alterations in lipid metabolism of the liver are the primary metabolic abnormalities which lead to hepatic steatosis (Videla et al., 2006).

The role of *n*-3 LC-PUFA as a potential therapeutic target in the pathogenesis of NAFLD has been demonstrated (Larter et al., 2008). The relationship between the *n*-6/*n*-3 PUFA ratio within the liver and the severity of steatosis has been reported. A higher *n*-6/*n*-3 LC-PUFA ratio within the liver of NAFLD patients may contribute to the development of fatty liver due to derangement in the capacity to regulate hepatic lipid metabolism (Araya et al., 2004; Vuppalanchi et al., 2007) resulting in fatty acid and triacylglycerol synthesis over oxidation (Vuppalanchi et al., 2007). Defective desaturation of PUFA due to inadequate intake of *n*-3 PUFA, along with a higher intake of *n*-6 PUFA, further enhances the contribution of desaturases inhibition in NAFLD (Patterson et al., 2012).

Moreover, oxidative stress may contribute to low hepatic LC-PUFA (Allard et al., 2008). The increase in lipid peroxidation associated with NAFLD, may contribute to the decrease in LC-PUFA, as they are particularly susceptible to lipid peroxidation (Allard et al., 2008). Thus, oxidative stress-dependent lipid peroxidation may represent an alternative mechanism to liver *n*-3 LC-PUFA depletion and eicosanoid derivatives in NAFLD (Videla et al., 2004). For example, LTB_4, an AA-derived eicosanoid, is involved in ROS production. The increased production of pro-inflammatory cytokines and eicosanoids, produced from *n*-6 PUFA metabolism, results in enhanced liver Kupffer cell production of inflammatory cytokines causing activation of NF-κB, further exacerbating systemic and hepatic insulin resistance with worsening inflammation and fibrosis (Byrne, 2010). Insulin resistance, as seen in NAFLD, may be related to the depletion of *n*-3 LC-PUFA because they are expected to modify membrane-mediated processes, such as insulin signaling (Lombardo and Chicco, 2006).

In summary, the depletion of *n*-3 LC-PUFA, the increase in *n*-6 PUFA, and the increase in *n*-6 PUFA-derived eicosanoid production within the liver all contribute to the development of NAFLD and related pathophysiologies, such as insulin resistance.

4.6. Osteoporosis

Abnormal lipid concentrations may contribute to the pathogenesis of osteoporosis (Ott, 2004). Evidence from animal studies suggests that excess precursor of PGE_2 leads to bone loss (Blanaru et al., 2004). Inflammation plays a distinct role in bone health through immune cell interactions and markers, such as C-reactive protein and IL-6, which are strongly associated with osteoporosis (Kelly et al., 2013). Osteoblastogenesis is inhibited by inflammation from exposure to high AA, but this effect is not seen with EPA (Shen et al., 2008). In this former study, the authors investigated if AA and EPA, at 40 μmol/L, had opposite effects on osteoblastogenesis and attempted to elucidate if COX-2 and inducible NO synthase played a role using the ST-2 cell line. Although the results were time dependent, overall, AA-treated cells produced more PGE_2 and had a higher value for COX-2 mRNA expression compared to EPA-treated cells. AA initially increased NO, but EPA resulted in higher NO values in the cells treated for a long time (Shen et al., 2008). The consensus from this and other comprehensive reviews is that high *n*-6 diets are associated with unfavorable bone outcomes whereas *n*-3 rich diets promote bone health in animal models (Li et al., 2010) and, to a limit extent, in humans (Kim and Ilich, 2011; Järvinen et al., 2012), according to life stages (Weiler, 2004). Moreover, when diets high in *n*-6 PUFA are modified to have a lower ratio of *n*-6/*n*-3 dietary fatty acids, bone mass is indeed enhanced (Blanaru et al., 2004).

One potential area of research concerning osteoporosis is to investigate the effect of dietary *n*-3 PUFA, or low ratios of *n*-6/*n*-3 PUFA in endocannabinoid metabolism, in particular, the indirect hypothalamic actions of endocannabinoids in bone during low-grade chronic inflammation (Kelly et al., 2013).

4.7. Rheumatoid Arthritis

Rheumatoid arthritis is a long-term, autoimmune disease characterized by flares of arthritis involving small joints and surrounding tissues, causing pain, swelling and impaired function. It features T-lymphocytes, macrophages, and plasma cells infiltration into the synovium, with the initiation of a chronic inflammatory state that involves the overproduction of eicosanoids and pro-inflammatory cytokines (Calder, 2008).

Previous studies have shown that in populations with traditionally high fish consumption, as the case of Greenland Eskimos, the percentage of AA in plasma phospholipids is about 20% of that found in industrialized countries (Bronsgeest-Schoute et al., 1981; Adam, 1992). The prevalence of rheumatoid arthritis in Eskimos populations is considerably lower than in inhabitants of Western communities (Kromann and Green, 1980). Studies have shown that AA-derived eicosanoids, PGE_2 and PGI_2, play a role in the pathogenesis of rheumatoid arthritis (Honda et al., 2006; Pulichino et al., 2006). PGI receptor-deficient mice ($IP^{-/-}$) subjected to collagen-induced arthritis showed a significant reduction in arthritic scores and a decline in IL-1β and IL-6 levels in the arthritic paws (Honda et al., 2006). The inhibition of both PGE receptors (EP_2 and EP_4) suppressed inflammatory events in collagen-induced arthritis. These results suggest that both PGE_2 and PGI_2 participate in rheumatoid arthritis. Moreover, the LTB_4 formation is increased by the inflammatory process and additionally augmented by the intake of COX inhibitors (Denzlinger et al., 1995).

The supplementation with *n*-3 PUFA has been demonstrated to modulate the activity of inflammatory factors that cause cartilage destruction during arthritis (Curtis et al., 2000; Calder, 2008). Fish oil studies in patients with rheumatoid arthritis have shown a decrease in LTB_4, as the most prominent feature of LOX inhibition (Kremer, 1996). This finding was confirmed later on by the lower LTB_4 production with fish oil supplementation in Adam et al., (2003) report.

In fact, clinical improvement and inhibition of eicosanoid formation were more pronounced in rheumatoid arthritis patients given EPA for 6-8 months, pointing towards a synergism between the low AA intake and the fish oil supplementation, and to the EPA/AA ratio as being decisive for clinical effectiveness (Adam et al., 2003). In the end, a reduction of dietary AA, from 171 mg to 49 mg, and of LA, from 12.7 g to 7.9 g, is able to reduce pain symptoms of rheumatoid arthritis by 14% and to improve the efficacy of fish oil supplementation, from 17% to 31-37%, in patients with rheumatoid arthritis (Adam et al., 2003). To conclude, limiting dietary intake of AA in Western diets, which are typically high in AA, ameliorates clinical signs of inflammation and appears to aid in symptoms of rheumatoid arthritis beyond augmenting the efficacy of fish oil supplementation (Adam et al., 2003).

4.8. Asthma

Several human lung cell types enzymatically oxygenate AA to generate diverse classes of eicosanoids. These metabolites are potent regulators of inflammatory reactions, smooth muscle tone, hemodynamic events, allergy and asthma, as well as many other important physiological processes in the lung (Levy, 2005).

In individuals with asthma, a bronchial allergen leads to enhanced production of PGD_2. This latter is a potent bronchoconstrictor, which is slightly more potent than its related prostaglandin $PGF_{2\alpha}$ and much more potent than histamine itself (Hardy et al., 1984; Bochenek et al., 2004). It is thought that signaling through the PGD_2 receptors mediates the pro-asthmatic effects of these prostaglandins (Gervais et al., 2001) and abolishing genetically these receptors is associated with reduced airway inflammation (Pettipher and Hansel, 2008). Thus, during asthmatic attacks in humans, PGD_2 is released in large amounts by mast cells (Arima and Fukuda, 2011) and promotes inflammation through activation of eosinophils (Hata and Breyer, 2004).

Even if AA eisosanoids appear to be preferentially pro-asthmatic (Pettipher and Hansel, 2008), the recent development of new lipoxins stable analogs that are topically and orally active should enable further investigation on lipoxins regulation in human respiratory illness. The diminished formation of these counter-regulatory mediators has been identified in aspirin-intolerant asthma (Sanak et al., 2000), severe steroid-dependent asthma (Bonnans et al., 2002), and cystic fibrosis airway (Karp et al., 2004) and may predispose to host to more severe inflammatory responses in the lung (Levy, 2005). This topic warrants further investigation.

CONCLUSION

At the end of this review, the following key points should be underlined. Dietary AA level is not associated, positively or negatively, with cancer risk, as demonstrated by several studies (Whelan and McEntee, 2004; Astorg,

2005). Likewise, dietary AA is not a risk factor for cardiovascular diseases, and may play a role in maintaining healthy cardiac (Harris et al., 2009) and cognitive function (Calder, 2007). For the epidemiology of heart disease, in particular, it may take many years for the high *n*-6 PUFA and low *n*-3 PUFA diet produce its negative health effects (Kelly et al., 2013). In line with this, a higher ratio of *n*-6/*n*-3 PUFA coupled with excess energy intake may be among the causes of the current obesity epidemic (Madsen et al., 2008) and associated metabolic disturbances. Notwithstanding, the safety of AA supplementation in patients suffering from cancer, inflammatory, neurodegenerative or other diseased states is unknown, and needs to be clarified by other results, especially those of ongoing prospective studies. In the field of nutritional lipid research, it may be helpful from hereafter, if authors reported the amount of PUFA in the table of nutrient intakes (Weiler, 2004), the complete fatty acid profiles of study diets and serum levels in animal studies (Kelly et al., 2013). In the case of *in vitro* experiments, the fatty acid profile of the sera used in cell cultures is also of considerable importance (Kelly et al., 2013).

It has generally been assumed that all AA-derived eicosanoids are pro-inflammatory and in fact, AA metabolism is a long recognized target for commonly used anti-inflammatory therapies. However, this is an over-simplification since as herein reviewed, some actions of eicosanoids are anti-inflammatory (e.g., PGE_2 inhibits production of some inflammatory cytokines) and it has been recently discovered that PGE_2 inhibits production of inflammatory leukotrienes and induces production of resolving LXA_4. Moreover, AA can be replaced, at least partially, into inflammatory cell phospholipids by EPA and DHA and therefore decrease the production of AA-derived mediators. EPA gives rise to an alternative family of eicosanoids (e.g., PGE_3), which frequently, but not always, have lower potency than those produced from AA. Recently, a new family of EPA- and DHA-derived lipid mediators called resolvins (E- and D-series) has been described. These molecules have potent anti-inflammatory and inflammation resolving properties in model systems. Thus, assumptions that: (a) all mediators formed from AA are pro-inflammatory; (b) eicosanoids produced from EPA are always less potent than those formed from AA; (c) EPA is the main, perhaps the only, anti-inflammatory *n*-3 PUFA; (d) anti-inflammatory actions of *n*-3 PUFA always relate to changes in synthesis of lipid mediators; have all been shown to be incorrect. It is now becoming clear that the actions of *n*-6 and *n*-3 PUFA and their derivatives on inflammation involve more mechanisms and are more complex than previously acknowledged (reviewed by Calder, 2009).

Thus, the identification of endogenous control mechanisms in the acute inflammatory response is of wide interest.

In addition, results from systematic reviews are promising in the sense that higher doses of *n*-3 PUFA seem to be safe and tolerable, regardless of the lack of large, long-term studies. Still to answer is left the question on the best study design to use in human *n*-3 PUFA trials (Kelly et al., 2013). Keeping in mind that the working definition of low-grade chronic inflammation emphasizes the importance of *n*-3 PUFA in both leading and treating low-grade chronic inflammation, the concept of using dietary factors, such as *n*-3 PUFA, to resolve inflammation is also new to nutritional studies, indicating that there may be a need to shift the focus from showing how *n*-3 PUFA can resolve inflammation and prevent low-grade chronic inflammation. The new pathways and bioactive maresins elegantly documented in Serhan et al., (2009) report, along with resolvins, protectins and related functional metabolome, might provide a new means to mark the impact of essential *n*-3 fatty acids in health and disease. Moreover, targeting lipoxins from AA metabolism, given their clinically relevant anti-inflammatory properties, might be worth investigating as a novel therapeutic approach for the resolution of inflammation.

ACKNOWLEDGMENT

Susana V. Martins is the recipient of Fundação para a Ciência e a Tecnologia (FCT) individual fellowship SFRH/BPD/63019/2009 and Paula A. Lopes is a researcher from the FCT program "Ciência 2008".

REFERENCES

Adam O. Immediate and long range effects of the uptake of increased amounts of arachidonic acid. *Clin. Investig.*, 1992; 70: 721-727.

Adam O., Beringer C., Kless T., Lemmen C., Adam A., Wiseman M., Adam P., Klimmek R., Forth W. Anti-inflammatory effects of a low arachidonic acid diet and fish oil in patients with rheumatoid arthritis. *Rheumatol. Int.,* 2003; 23: 27-36.

Aiello R. J., Bourassa P. A., Lindsey S., Weng W., Freeman A., Showell H. J. Leukotriene B4 receptor antagonism reduces monocytic foam cells in mice. *Arterioscler. Thromb. Vasc. Biol.*, 2002; 22: 443-449.

Allard J. P., Aghdassi E., Mohammed S., Raman M., Avand G., Arendt B. M., Jalali P., Kandasamy T., Prayitno N., Sherman M., Guindi M., Ma D. W., Heathcote J. E. Nutritional assessment and hepatic fatty acid composition in non-alcoholic fatty liver disease (NAFLD): a cross-sectional study. *J. Hepatol.,* 2008; 48: 300-307.

Almer G., Guegan C., Teismann P., Naini A., Rosoklija G., Hays A. P., Chen C., Przedborski S. Increased expression of the proinflammatory enzyme cyclooxygenase-2 in amyotrophic lateral sclerosis. *Ann. Neurol.,* 2001; 49: 176-185.

Amtul Z., Uhrig M., Wang L., Rozmahel R. F., Beyreuther K. Detrimental effects of arachidonic acid and its metabolites in cellular and mouse models of Alzheimer's disease: structural insight. *Neurobiol. Aging,* 2012; 33: 831.e21-831.e31.

Araya J., Rodrigo R., Videla L. A., Thielemann L., Orellana M., Pettinelli P., Poniachik J. Increase in long-chain polyunsaturated fatty acid n-6/n-3 ratio in relation to hepatic steatosis in patients with non-alcoholic fatty liver disease. *Clin. Sci.,* 2004; 106: 635-643.

Ariel A., Serhan C. N. Resolvins and protectins in the termination program of acute inflammation. *Trends Immunol.,* 2007; 28: 176-183.

Arima M., Fukuda T. Prostaglandin D2 and TH2 Inflammation in the pathogenesis of bronchial asthma. *Korean J. Intern. Med.*, 2011; 26: 8-18.

Arita M., Bianchini F., Aliberti J., Sher A., Chiang N., Hong S., Yang R., Petasis N. A., Serhan C. N. Stereochemical assignment, antiinflammatory properties and receptor for the omega-3 lipid mediator resolving E1. *J. Exp. Med.,* 2005; 201: 713-722.

Aro A., Männistö S., Salminen I., Ovaskainen M. L., Kataja V., Uusitupa M. Inverse association between dietary and serum conjugated linoleic acid and risk of breast cancer in postmenopausal women. *Nutr. Cancer*, 2000; 38: 151-157.

Arterburn L. M., Boswell K. D., Lawlor T., Cifone M. A., Murli H., Kyle D. J. In vitro genotoxicity testing of ARASCO and DHASCO oils. *Food Chem. Toxicol.*, 2000a; 38: 971-976.

Arterburn L. M., Boswell K. D., Henwood S. M., Kyle D. J. A developmental safety study in rats using DHA- and ARA-rich single-cell oils. *Food Chem. Toxicol.,* 2000b; 38: 763-771.

Artwohl M., Lindenmair A., Sexl V., Maier C., Rainer G., Freudenthaler A., Huttary N., Wolzt M., Nowotny P., Luger A., Baumgartner-Parzer S. M. Different mechanisms of saturated versus polyunsaturated FFA-induced apoptosis in human endothelial cells. *J. Lipid Res.,* 2008; 49: 2627-2640.

Astorg P. Dietary fatty acids and colorectal and prostate cancers: epidemiological studies. *Bull. Cancer*, 2005; 92: 670-684.

Azrad M., Turgeon C., Demark-Wahnefried W. Current evidence linking polyunsaturated fatty acids with cancer risk and progression. *Front. Oncol.,* 2013; 3: 224.

Bachi A. L., Kim F. J., Nonogaki S., Carneiro C. R., Lopes J. D., Jasiulionis M. G., Correa M. Leukotriene B4 creates a favorable microenvironment for murine melanoma growth. *Mol. Cancer Res.*, 2009; 7: 1417-1424.

Bagga D., Wang L., Farias-Eisner R., Glaspy J. A., Reddy S. T. Differential effects of prostaglandin derived from ω-6 and ω-3 polyunsaturated fatty acids on Cox-2 expression and IL-6 secretion. *Proc. Natl. Acad. Sci. U S A,* 2003; 100: 1751-1756.

Balsinde J., Bianco I. D., Ackermann E. J., Conde-Frieboes K., Dennis E. A. Inhibition of calcium-independent phospholipase A2 prevents arachidonic acid incorporation and phospholipid remodeling in P388D1 macrophages. *Proc. Natl. Acad. Sci. U S A,* 1995; 92: 8527-8531.

Balsinde J., Winstead M. V., Dennis E. A. Phospholipase A(2) regulation of arachidonic acid mobilization. *FEBS Lett.,* 2002; 531: 2-6.

Bannenberg G., Serhan C. N. Specialized pro-resolving lipid mediators in the inflammatory response: An update. *Biochim. Biophys. Acta*, 2010; 1801: 1260-1273.

Bazan N. G., Colangelo V., Lukiw W. J. Prostaglandins and other lipid mediators in Alzheimer's disease. *Prostaglandins Other Lipid Mediat.,* 2002; 68-69: 197-210.

Bazinet R. P. Is the brain arachidonic acid cascade a common target of drugs used to manage bipolar disorder? *Biochem. Soc. Trans.*, 2009; 37: 1104-1109.

Bederska-Lojewska D., Orczewska-Dudek S., Pieszka M. Metabolism of arachidonic acid, its concentration in animal products and influence on inflammatory processes in the human body: a review. *Ann. Anim. Sci.,* 2013; 13: 177-194.

Bensinger S. J., Tontonoz P. Integration of metabolism and inflammation by lipid-activated nuclear receptors. *Nature*, 2008; 454: 470-477.

Birch E. E., Garfield S., Hoffman D. R., Uauy R., Birch D. G. A randomized controlled trial of early dietary supply of long-chain polyunsaturated fatty acids and mental development in term infants. *Dev. Med. Child Neurol.,* 2000; 42: 174-181.

Birch E. E., Garfield S., Castañeda Y., Hughbanks-Wheaton D., Uauy R., Hoffman D. R., Birch D. G. Visual acuity and cognitive outcomes at 4 years of age in a double-blind, randomized trial of long-chain polyunsaturated fatty acid-supplemented infant formula. *Early Hum. Dev.*, 2007; 83: 279-284.

Blanaru J. L., Kohut J. R., Fitzpatrick-Wong S. C., Weiler H. A. Dose response of bone mass to dietary arachidonic acid in piglets fed cow milk-based formula. *Am. J. Clin. Nutr.*, 2004; 79: 139-147.

Bocca C., Bozzo F., Martinasso G., Canuto R. A., Miglietta A. Involvement of PPARα in the growth inhibitory effect of arachidonic acid on breast cancer cells. *Br. J. Nutr.*, 2008; 100: 739-750.

Bochenek G., Nizankowska E., Gielicz A., Swierczynska M., Szczeklik A. Plasma 9α,11β-PGF2, a PGD2 metabolite, as a sensitive marker of mast cell activation by allergen in bronchial asthma. *Thorax,* 2004; 59: 459-464.

Bogatcheva N. V., Sergeeva M. G., Dudek S. M., Verin A. D. Arachidonic acid cascade in endothelial pathobiology. *Microvasc. Res.*, 2005; 69: 107-127.

Bondesen B. A., Mills S. T., Kegley K. M., Pavlath G. K. The COX-2 pathway is essential during early stages of skeletal muscle regeneration. *Am. J. Physiol. Cell Physiol.*, 2004; 287: C475-483.

Bonnans C., Vachier I., Chavis C., Godard P., Bousquet J., Chanez P. Lipoxins are potential endogenous antiinflammatory mediators in asthma. *Am. J. Respir. Crit. Care Med.,* 2002; 165: 1531-1535.

Bosma-den Boer M. M., van Wetten M. L., Pruimboom L. Chronic inflammatory diseases are stimulated by current lifestyle: how diet, stress levels and medication prevent our body from recovering. *Nutr. Metab.*, 2012; 9: 32.

Borkman M., Storlien L. H., Pan D. A., Jenkins A. B., Chisholm D. J., Campbell L. V. The relation between insulin sensitivity and the fatty-acid composition of skeletal-muscle phospholipids. *New England J. Med.*, 1993; 328: 238-244.

Brash A. R. Arachidonic acid as a bioactive molecule. *J. Clin. Invest.,* 2001; 107: 1339-1345.

Breckenridge W. C., Yeung S. K. F., Kuksis A. Biosynthesis of triacylglycerols by rat intestinal mucosa in vivo. *Can. J. Biochem.,* 1976; 54: 145-152.

Brenna J. T., Varamini B., Jensen R. G., Diersen-Schade D. A., Boettcher J. A., Arterburn L. M. Docosahexaenoic and arachidonic acid concentrations in human breast milk worldwide. *Am. J. Clin. Nutr.,* 2007; 85: 1457-1464.

Breyer R. M., Bagdassarian C. K., Myers S. A., Breyer M. D. Prostanoid receptors: subtypes and signalling. *Annu. Rev. Pharmacol. Toxicol.*, 2001; 41: 661-690.

Bronsgeest-Schoute H. C., van Gent C. M., Luten J. B., Ruiter A. The effect of various intakes of omega 3 fatty acids on the blood lipid composition in healthy human subjects. *Am. J. Clin. Nutr.*, 1981; 34: 1752-1757.

Burke J. E., Dennis E. A. Phospholipase A2 structure/function, mechanism, and signalling. *J. Lipid Res.,* 2009; 50: S237-S242.

Byrne C. D., Olufadi R., Bruce K. D., Cagampang F. R., Ahmed M. H. Metabolic disturbances in non-alcoholic fatty liver disease. *Clin. Sci.,* 2009; 116: 539-564.

Byrne C. D. Fatty liver: Role of inflammation and fatty acid nutrition. *Prostaglandins Leukot. Essent. Fatty Acids*, 2010; 82: 265-271.

Calder P. C. Polyunsaturated fatty acids, inflammation, and immunity. *Lipids,* 2001; 36: 1007-1024.

Calder P. C. n-3 polyunsaturated fatty acids, inflammation, and inflammatory diseases. *Am. J. Clin. Nutr.*, 2006; 83: 1505S-1519S.

Calder P. C. Dietary arachidonic acid: harmful, harmless or helpful. *Br. J. Nutr.,* 2007; 98: 451-453.

Calder P. C. PUFA, inflammatory processes and rheumatoid arthritis. *Proc. Nutr. Soc.*, 2008; 67: 409-418.

Calder P. C. Polyunsaturated fatty acids and inflammatory processes: new twists in an old tale. *Biochimie.,* 2009; 91: 791-795.

Calder P. C. Long-chain fatty acids and inflammation. *Proc. Nutr. Soc.,* 2012; 71: 284-289.

Cancello R., Clement K. Is obesity an inflammatory illness? Role of low-grade inflammation and macrophage infiltration in human white adipose tissue. *BJOG*, 2006; 113: 1141-1147.

Capdevila J. H., Falck J. R., Harris R. C. Cytochrome P450 and arachidonic acid bioactivation: molecular and functional properties of the arachidonate monooxygenase. *J. Lipid Res.,* 2000; 41: 163-181.

Cathcart M. C., Lysaght J., Pidgeon G. P. Eicosanoid signaling pathways in the development and progression of colorectal cancer: novel approaches for prevention/intervention. *Cancer Metastasis Rev.,* 2011; 30: 363-368.

Chan T. A., Morin P. J., Vogelstein B., Kinzler K. W. Mechanisms underlying nonsteroidal antiinflammatory drug-mediated apoptosis. *Proc. Natl. Acad. Sci. U S A,* 1998; 95: 681-686.

Chang K., Chang W. H. Pulsed electromagnetic fields prevent osteoporosis in an ovariectomized female rat model: A prostaglandin E2-associated process. *Bioelectromagnetics,* 2003; 24: 189-198.

Chavarro J. E., Stampfer M. J., Li H., Campos H., Kurth T., Ma J. A prospective study of polyunsatutated fatty acid levels in blood and prostate cancer risk. *Cancer Epidemiol. Biomarkers Prev.,* 2007; 16: 1364-1370.

Chen Y., Hao H., He S., Cai L., Li Y., Hu S., Ye D., Hoidal J., Wu P., Chen X. Lipoxin A4 and its analogue suppress the tumor growth of transplanted H22 in mice: the role of antiangiogenesis. *Mol. Cancer Ther.,* 2011; 9: 2164-2174.

Chiang N., Serhan C. N., Dahlén S-E., Drazen J. M., Hay D. W. P., Rovati G. E., Shimizu T., Yokomizo T., Brink C. The lipoxin receptor ALX: Potent ligand-specific and stereoselective actions in vivo. *Pharmacol. Rev.,* 2006; 58: 463-487.

Chilton F. H., Fonteh A. N., Surette M. E., Triggiani M., Winkler J. D. Control of arachidonate levels within inflammatory cells. *Biochim. Biophys. Acta.,* 1996; 1299: 1-15.

Christie M. J., Vaughan C. W., Ingram S. L. Opioids, NSAIDs and 5-lipoxygenase inhibitors act synergistically in brain via arachidonic acid metabolism. *Inflamm. Res.,* 1999; 48: 1-4.

Chu H. W., Balzar S., Westcott J. Y., Trudeau J. B., Sun Y., Conrad D. J., Wenzel S. E. Expression and activation of 15-lipoxygenase pathway in severe asthma: relationship to eosinophilic phenotype and collagen deposition. *Clin. Exp. Allergy,* 2002; 32: 1558-1565.

Cinti S. Histopathology of the obese adipose organ. *Ital. J. Anat. Embriol.,* 2011; 116: 44.

Clandinin M. T., Van Aerde J. E., Merkel K. L., Harris C. L., Springer M. A., Hansen J. W., Diersen-Schade D. A. Growth and development of preterm infants fed infant formulas containing docosahexaenoic acid and arachidonic acid. *J. Pediatr.,* 2005; 146: 461-468.

Claria J., Dalli J., Yacoubian S., Gao F., Serhan C. N. Resolvin D1 and resolving D2 govern local inflammatory tone in obese fat. *J. Immunol.,* 2012; 189: 2597-2605.

Cordain L., Eaton S. B., Sebastian A., Mann N., Lindeberg S., Watkins B. A., O'Keefe J. H., Brand-Miller J. Origins and evolution of the Western diet: health implications for the 21st century. *Am. J. Clin. Nutr.,* 2005; 81: 341-354.

Crowe F. L., Allen N. E., Appleby P. N., Overvad K., Aardestrup I. V., Johnsen N. F., Tjønneland A., Linseisen J., Kaaks R., Boeing H., Kröger

J., Trichopoulou A., Zavitsanou A., Trichopoulos D., Sacerdote C., Palli D., Tumino R., Agnoli C., Kiemeney L. A., Bueno-de-Mesquita H. B., Chirlaque M. D., Ardanaz E., Larrañaga N., Quirós J. R., Sánchez M. J., González C. A., Stattin P., Hallmans G., Bingham S., Khaw K. T., Rinaldi S., Slimani N., Jenab M., Riboli E., Key T. J. Fatty acid composition of plasma phospholipids and risk of prostate cancer in a case-control analysis nested within the European Prospective Investigation into Cancer and Nutrition. *Am. J. Clin. Nutr.*, 2008; 88: 1353-1363.

Curtis C. L., Hughes C. E., Flannery C. R., Little C. B., Harwood J. L., Caterson B. *n*-3 fatty acids specifically modulate catabolic factors involved in articular cartilage degradation. *J. Biol. Chem.*, 2000; 275: 721-724.

Dalli J., Winkler J. W., Colas R. A., Arnardottir H., Cheng C. Y., Chiang N., Petasis N. A., Serhan C. N. Resolvin D3 and aspirin-triggered resolving D3 are potent immunoresolvents. *Chem. Biol.*, 2013; 20: 188-201.

Darios F., Davletov B. Omega-3 and omega-6 fatty acids stimulate cell membrane expansion by acting on syntaxin 3. *Nature,* 2006; 440: 813-817.

Denzlinger C., Kless T., Sagebiel-Kohler S., Lemmen C., Jacob K., Wilmanns W., Adam O. Modulation of the endogenous leukotriene production by fish oil and vitamin E. *J. Lipid Mediat. Cell Signal*, 1995; 11: 119-132.

Dunstan J. A., Mitoulas L. R., Dixon G., Doherty D. A., Hartmann P. E., Simmer K., Prescott S. L. The effects of fish oil supplementation in pregnancy on breast milk fatty acid composition over the course of lactation: a randomized controlled trial. *Pediatr. Res.,* 2007; 62: 689-694.

Elmadfa I., Kornsteiner M. Fats and fatty acid requirements for adults. *Ann. Nutr. Metab.,* 2009; 55: 56-75.

Erickson S. Nonalcoholic fatty liver disease. *J. Lipid Res.,* 2009; 50: S412-S416.

Faas F. H., Dang A. Q., White J., Schaefer R. F., Johnson D. E. Decreased prostatic arachidonic acid in human prostatic carcinoma. *BJU Int.,* 2003; 92: 551-554.

Fang K. M., Chang W. L., Wang S. M., Su M. J., Wu M. L. Arachidonic acid induces both Na^+ and Ca^{2+} entry resulting in apoptosis. *J. Neurochem.,* 2008; 104: 1177-1189.

Farooqui A. A., Ong W. Y., Horrocks L. A. Inhibitors of brain phospholipase A2 activity: their neuropharmacological effects and therapeutic importance for the treatment of neurologic disorders. *Pharmacol. Rev.,* 2006; 58: 591-620.

Ferrucci L., Cherubini A., Bandinelli S., Bartali B., Corsi A., Lauretani F., Martin A., Andres-Lacueva C., Senin U., Guralnik J. M. Relationship of plasma polyunsaturated fatty acids to circulating inflammatory markers. *J. Clin. Endocrinol. Metab.*, 2006; 91: 439-446.

Field C. J., Thomson C. A., Van Aerde J. E., Parrott A., Euler A., Lien E., Clandinin M. T. Lower proportion of CD45R0+ cells and deficient interleukin-10 production by formula-fed infants, compared with human-fed, is corrected with supplementation of long-chain polyunsaturated fatty acids. *J. Pediatr. Gastroenterol. Nutr.*, 2000; 31: 291-299.

Fiorio Pla A., Munaron L. Calcium influx, arachidonic acid, and control of endothelial cell proliferation. *Cell Calcium*, 2001; 30: 235-244.

Flachs P., Horakova O., Brauner P., Rossmeisl M., Pecina P., Franssen-van Hal N., Ruzickova J., Sponarova J., Drahota Z., Vlcek C., Keijer J., Houstek J., Kopecky J. Polyunsaturated fatty acids of marine origin upregulate mitochondrial biogenesis and induce beta-oxidation in white fat. *Diabetologia,* 2005; 48: 2365-2375.

Flachs P., Rossmeisl M., Bryhn M., Kopecky J. Cellular and molecular effects of n-3 polyunsaturated fatty acids on adipose tissue biology and metabolism. *Clin. Sci.,* 2009; 116: 1-16.

Folco G., Murphy R. C. Eicosanoid transcellular biosynthesis: from cell-cell interactions to in vivo tissue responses. *Pharmacol. Rev.,* 2006; 58: 375-388.

Fukaya T., Gondaira T., Kashiyae Y., Kotani S., Ishikura Y., Fujikawa S., Kiso Y., Sakakibara M. Arachidonic acid preserves hippocampal neuron membrane fluidity in senescent rats. *Neurobiol. Aging,* 2007; 28: 1179-1186.

Furuyashiki T., Narumiya S. Stress responses: the contribution of prostaglandin E2 and its receptors. *Nat. Rev. Endocrinol.*, 2011; 7: 163-175.

Garland M., Sacks F. M., Colditz G. A., Rimm E. B., Sampson L. A., Willett W. C., Hunter D. J. The relation between dietary intake and adipose tissue composition of selected fatty acids in US women. *Am. J. Clin. Nutr.,* 1998; 67: 25-30.

Gervais F. G., Cruz R. P. G., Chateauneuf A., Gale S., Sawyer N., Nantel F., Metters K. M., O'Neill G. P. Selective modulation of chemokinesis, degranulation, and apoptosis in eosinophils through the PGD2 receptors CRTH2 and DP. *J. Allergy Clin. Immunol.*, 2001; 108: 982-988.

Gewirtz A. T., Collier-Hyams L. S., Young A. N., Kucharzik T., Guilford W. J., Parkinson J. F., Williams I. R., Neish A. S., Madara J. L. Lipoxin A4 analogs attenuate induction of intestinal epithelial proinflammatory gene expression and reduce the severity of dextran sodium sulfate-induced colitis. *J. Immunol.*, 2002; 168: 5260-5267.

Ghadimi R., Kuriki K., Tsuge S., Takeda E., Imaeda N., Suzuki S., Sawai A., Takekuma K., Hosono A., Tokudome Y., Goto C., Esfandiary I., Nomura H., Tokudome S. Serum concentrations of fatty acids and colorectal adenoma risk: a case-control study in Japan. *Asian Pac. J. Cancer Prev.*, 2008; 9: 111-118.

Ghosh J., Myers C. E. Arachidonic acid stimulates prostate cancer cell growth: critical role of 5-lipoxygenase. *Biochem. Biophys. Res. Commun.*, 1997; 235: 418-423.

Giudetti A. M., Cagnazzo R. Beneficial effects of n-3 PUFA on chronic airway inflammatory diseases. *Prostaglandins Other Lipid Mediat.*, 2012; 99: 57-67.

Godson C., Mitchell S., Harvey K., Petasis N. A., Hogg N., Brady H. R. Cutting edge: lipoxins rapidly stimulate nonphlogistic phagocytosis of apoptotic neutrophils by monocyte-derived macrophages. *J. Immunol.*, 2000; 164: 1663-1667.

Goldberg A. L., Baracos V., Rodemann P., Waxman L., Dinarello C. Control of protein degradation in muscle by prostaglandins, Ca2+, and leukocytic pyrogen (interleukin 1). *Fed. Proc.*,1984; 43: 1301-1306.

Griswold D. E., Adams J. L. Constitutive cyclooxygenase (COX-1) and inducible cyclooxygenase (COX-2): rationale for selective inhibition and progress to date. *Med. Res. Rev.*, 1996; 16: 181-206.

Haeggström J. Z., Rinaldo-Matthis A., Wheelock C. E., Wetterholm A. Advances in eicosanoid research, novel therapeutic implications. *Biochem. Bioph. Res. Commun.*, 2010; 396: 135-139.

Hagi A., Nakayama S., Shinzaki W., Haji S., Ohyanagi H. Effects of the omega-6:omega-3 fatty acid ratio of fat emulsions on the fatty acid composition in cell membranes and the anti-inflammatory action. *J. Parenter. Enteral. Nutr.*, 2010; 34: 263-270.

Hardy C. C., Robinson C., Tattersfield A. E., Holgate S. T. The bronchoconstrictor effect of inhaled prostaglandin D2 in normal and asthmatic men. *N. Engl. J. Med.*, 1984; 311: 209-213.

Harizi H., Corcuff J. B., Gualde N. Arachidonic acid-derived eicosanoids: roles in biology and immunopathology. *Trends Mol. Med.*, 2008; 14: 461-469.

Harris S. G., Padilla J., Koumas L., Ray D., Phipps R. P. Prostaglandins as modulators of immunity. *Trends Immunol.*, 2002; 23: 144-150.

Harris R. B., Foote J. A., Hakim I. A., Bronson D. L., Alberts D. S. Fatty acid composition of red blood cell membranes and risk of squamous cell carcinoma of the skin. *Cancer Epidemiol. Biomarkers Prev.,* 2005; 14: 906-912.

Harris W. S., Mozaffarian D., Rimm E., Kris-Etherton P., Rudel L. L., Appel L. J., Engler M. M., Engler M. B., Sacks F. Omega-6 fatty acids and risk for cardiovascular disease: a science advisory from the American Heart Association Nutrition Subcommittee of the Council on Nutrition, Physical Activity, and Metabolism; Council on Cardiovascular Nursing; and Council on Epidemiology and Prevention. *Circulation,* 2009; 119: 902-907.

Hata A. N., Breyer R. M. Pharmacology and signaling of prostaglandin receptors: multiple roles in inflammation and immune modulation. *Pharmacol. Ther.,* 2004; 103: 147-166.

Hayaishi O., Matsumura H. Prostaglandins and sleep. *Adv. Neuroimmunol.,* 1995; 5: 211-216.

Hempenius R. A., Van Delft J. M., Prinsen M., Lina B. A. Preliminary safety assessment of an arachidonic acid-enriched oil derived from *Mortierella alpina*: summary of toxicological data. *Food Chem. Toxicol.*, 1997; 35: 573-581.

Hempenius R. A., Lina B. A., Haggitt R. C. Evaluation of a subchronic (13-week) oral toxicity study, preceded by an in utero exposure phase, with arachidonic acid oil derived from *Mortierella alpina* in rats. *Food Chem. Toxicol.,* 2000; 38: 127-139.

Herrera B. S., Ohira T., Gao L., Omori K., Yang R., Zhu M., Muscara M. N., Serhan C. N., Van Dyke T. E., Gyurko R. An endogenous regulator of inflammation, resolving E1, modulates osteoclast differentiation and bone resorption. *Br. J. Pharmacol.*, 2008; 155: 1214-1223.

His L. C., Wilson L. C., Eling T. E. Opposing effects of 15-lipoxygenase-1 and -2 metabolites on MAPK signaling in prostate. *J. Biol. Chem.,* 2002; 277: 40549-40556.

Honda T., Segi-Nishida E., Miyachi Y., Narumiya S. Prostacyclin-IP signaling and prostaglandin E2-EP2/EP4 signaling both mediate joint inflammation in mouse collagen-induced arthritis. *J. Exp. Med.,* 2006; 203: 325-335.

Horsley V., Pavlath G. K. Prostaglandin $F_{2\alpha}$ stimulates growth of skeletal muscle cells via an NFATC2-dependent pathway. *J. Cell Biol.*, 2003; 161: 111-118.

Hughes C. L., Dhiman T. R. Dietary compounds in relation to dietary diversity and human health. *J. Medicinal Food*, 2002; 5: 51-68.

Imai N., Kawabe M., Tamano S., Doi Y., Nakashima H., Suguro M., Numano T., Hara T., Hagiwara A., Furukawa F., Kaneda Y., Tateishi N., Fujii W., Kawashima H., Shibata H., Sakakibara Y. Arachidonate-enriched triglyceride oil does not promote tumor development in a rat medium-term multi-organ carcinogenesis model. *Food Chem. Toxicol.*, 2012; 50: 2780-2791.

Ishikura Y., Ikeda G., Akimoto K., Hata M., Kusumoto A., Kidokoro A., Kontani M., Kawashima H., Kiso Y., Koga Y. Arachidonic acid supplementation decreases P300 latency and increases P300 amplitude of event-related potentials in healthy elderly men. *Neuropsychobiology,* 2009; 60: 73-79.

Jakowec M. W., Petzinger G. M. 1-Methyl-4-phenyl-1,2,3,6-tetrahydropyridine-lesioned model of parkinson's disease, with emphasis on mice and nonhuman primates. *Comp. Med.*, 2004; 54: 497-513.

Järvinen R., Tuppurainen M., Erkkilä A. T., Penttinen P., Kärkkäinen M., Salovaara K., Jurvelin J. S., Kröger H. Associations of dietary polyunsaturated fatty acids with bone mineral density in elderly women. *Eur. J. Clin. Nutr.,* 2012; 66: 496-503.

Jayadev S., Linardic C. M., Hannun Y. A. Identification of arachidonic acid as a mediator of sphingomyelin hydrolysis in response to tumor necrosis factor alpha. *J. Biol. Chem.,* 1994; 269: 5757-5763.

Jensen R. G. Lipids in human milk. *Lipids,* 1999; 34: 1243-1271.

Jonnalagadda S. S., Egan S. K., Heimbach J. T., Harris S. S., Kris-Etherton P. M. Fatty acid consumption patterns of Americans: 1987–1988 USDA Nationwide Food Consumption Survey. *Nutr. Res.,* 1995; 15: 1767-1781.

Jubelirer S. J., Russel F., Vaillancourt R., Deykin D. Platelet arachidonic acid metabolism and platelet function in ten patients with chronic myelogenous leucemia. *Blood,* 1980; 56: 728-731.

Kaaman M., Rydén M., Axelsson T., Nordström E., Sicard A., Bouloumié A., Langin D., Arner P., Dahlman I. ALOX5AP expression, but no gene haplotypes, is associated with obesity and insulin resistance. *Int. J. Obes.*, 2006; 30: 447-452.

Kakutani S., Ishikura Y., Tateishi N., Horikawa C., Tokuda H., Kontani M., Kawashima H., Sakakibara Y., Kiso Y., Shibata H., Morita I. Supplementation of arachidonic acid-enriched oil increases arachidonic acid contents in plasma phospholipids, but does not increase their metabolites and clinical parameters in Japanese healthy elderly individuals: a randomized controlled study. *Lipids Health Dis.,* 2011; 10: 241.

Karp C. L., Flick L. M., Park K. W., Softic S., Greer T. M., Keledjian R., Yang R., Uddin J., Guggino W. B., Atabani S. F., Belkaid Y., Xu Y., Whitsett J. A., Accurso F. J., Wills-Karp M., Petasis N. A. Defective lipoxin-mediated anti-inflammatory activity in the cystic fibrosis airway. *Nat. Immunol.,* 2004; 5: 388-392.

Kaufmann W. E., Worley P. F., Pegg J., Bremer M., Isakson P. COX-2, a synaptically induced enzyme, is expressed by excitatory neurons at postsynaptic sites in rat cerebral cortex. *Proc. Natl. Acad. Sci. U S A,* 1996; 93: 2317-2321.

Kawabata T., Hirota S., Hirayama T., Adachi N., Hagiwara C., Iwama N., Kamachi K., Araki E., Kawashima H., Kiso Y. Age-related changes of dietary intake and blood eicosapentaenoic acid, docosahexaenoic acid, and arachidonic acid levels in Japanese men and women. *Prostaglandins Leukot. Essent. Fatty Acids,* 2011; 84: 131-137.

Kawasaki A., Han M. H., Wei J. Y., Hirata K., Otori Y., Barnstable C. J. Protective effects of arachidonic acid on glutamate neurotoxicity in rat retinal ganglion cells. *Invest. Ophthalmol. Vis. Sci.*, 2002; 43: 1835-1842.

Keane D. C., Takahashi H. K., Dhayal S., Morgan N. G., Curi R., Newsholme P. Arachidonic acid actions on functional integrity and attenuation of the negative effects of palmitic acid in a clonal pancreatic β-cell line. *Clin. Sci.,* 2011; 120: 195-206.

Kelley D. S., Taylor P. C., Nelson G. J., Schmidt P. C., Mackey B. E., Kyle D. Effects of dietary arachidonic acid on human immune response. *Lipids,* 1997; 32: 449-456.

Kelley D. S., Taylor P. C., Nelson G. J., MacKey B. E. Arachidonic acid supplementation enhances synthesis of eicosanoids without suppressing immune functions in young healthy men. *Lipids,* 1998; 33: 125-130.

Kelly O. J., Gilman J. C., Kim Y., Ilich Jz. Long-chain polyunsaturated fatty acids may mutually benefit both obesity and osteoporosis. *Nutr. Res.,* 2013; 33: 521-533.

Khairallah R. J., Kim J., O'Shea K. M., O'Connell K. A., Brown B. H., Galvao T., Daneault C., Rosiers C., Polster B., Hoppel C., Stanley W. C. Improved mitochondrial function with diet-induced increase in either docosahexaenoic acid or arachidonic acid in membrane phospholipids. *PLoS One,* 2012; 7: e34402.

Kim H. K., Della-Fera M., Lin J., Baile C. A. Docosahexaenoic acid inhibits adipocyte differentiation and induces apoptosis in 3T3-L1 adipocytes. *J. Nutr.,* 2006; 136: 2965-2969.

Kim Y., Ilich J. Z. Implications of dietary α-linolenic acid in bone health. *Nutrition,* 2011; 27: 1101-1107.

Kobayashi J., Yokoyama S., Kitamura S. Eicosapentaenoic acid modulates arachidonic acid metabolism in rat alveolar macrophages. *Prostaglandins Leukot. Essent. Fatty Acids,* 1995; 52: 259-262.

Kobayashi M., Sasaki S., Kawabata T., Hasegawa K., Tsugane S., JPHC. Validity of a self-administered food frequency questionnaire used in the 5-year follow-up survey of the JPHC Study Cohort I to assess fatty acid intake: comparison with dietary records and serum phospholipid level. *J. Epidemiol.,* 2003; 13: S64-S81.

Koletzko B., Decsi T., Demmelmair H. Arachidonic acid supply and metabolism in human infants born at full term. *Lipids,* 1996; 31: 79-83.

Koskelo E-K., Boswell K., Carl L., Lanoue S., Kelly C., Kyle D. High level of dietary arachidonic acid triglyceride exhibit no subchronic toxicity in rats. *Lipids,* 1997; 32: 397-405.

Kotani S., Nakazawa H., Tokimasa T., Akimoto K., Kawashima H., Toyoda-Ono Y., Kiso Y., Okaichi H., Sakakibara M. Synaptic plasticity preserved with arachidonic acid diet in aged rats. *Neurosci. Res.*, 2003; 46: 453-461.

Kotani S., Sakaguchi E., Warashina S., Matsukawa N., Ishikura Y., Kiso Y., Sakakibara M., Yoshimoto T., Guo J., Yamashima T. Dietary supplementation of arachidonic and docosahexaenoic acids improves cognitive dysfunction. *Neurosci. Res.,* 2006; 56: 159-164.

Kremer M. Effects of modulation of inflammatory and immune parameters in patients with rheumatic and inflammatory disease receiving dietary supplementation of *n*-3 and *n*-6 fatty acids. *Lipids,* 1996; 31: S243-S247.

Kremmyda L-S., Tvrzicka E., Stankova B., Zak A. Fatty acids as biocompounds: their role in human metabolism, health and disease – a review. Part 2: Fatty acid physiological roles and applications in human health and disease. *Biomed. Pap. Med. Fac. Univ. Palacky Olomouc. Czech Repub.,* 2011; 155: 195-218.

Kromann N., Green A. Epidemiological studies in the Upernavik district, Greenland. Incidence of some chronic diseases 1950-1974. *Acta Med. Scand.,* 1980; 208: 401-406.

Kuhn H., Chaitidis P., Roffeis J., Walther M. Arachidonic Acid metabolites in the cardiovascular system: the role of lipoxygenase isoforms in atherogenesis with particular emphasis on vascular remodeling. *J. Cardiovasc. Pharmacol.,* 2007; 50: 609-620.

Kuriki K., Wakai K., Hirose K., Matsuo K., Ito H., Suzuki T., Saito T., Kanemitsu Y., Hirai T., Kato T., Tatematsu M., Tajima K. Risk of colorectal cancer is linked to erythrocyte compositions of fatty acids as biomarkers for dietary intakes of fish, fat, and fatty acids. *Cancer Epidemiol. Biomarkers Prev.,* 2006; 15: 1791-1798.

Kusumoto A., Ishikura Y., Kawashima H., Kiso Y., Takai S., Miyazaki M. Effects of arachidonate-enriched triacylglycerol supplementation on serum fatty acids and platelet aggregation in healthy male subjects with a fish diet. *Br. J. Nutr.,* 2007; 98: 626-635.

Larter C. Z., Yeh M. M., Cheng J., Williams J., Brown S., dela Pena A., Bell-Anderson K. S., Farrell G. C. Activation of peroxisome proliferator-activated receptor alpha by dietary fish oil attenuates steatosis, but does not prevent experimental steatohepatitis because of hepatic lipoperoxide accumulation. *J. Gastroenterol. Hepatol.,* 2008; 23: 267-275.

Laschober G. T., Brunauer R., Jamnig A., Singh S., Hafen U., Fehrer C., Kloss F., Gassner R., Lepperdinger G. Age-specific changes of mesenchymal stem cells are paralleled by upregulation of CD106 expression as a response to an inflammatory environment. *Rejuvenation Res.*, 2011; 14: 119-131.

Lawson J. A., Rokach J., FitzGerald G. A. Isoprostanes: formation, analysis and use as indices of lipid peroxidation in vivo. *J. Biol. Chem.,* 1999; 274: 24441-24444.

Lerea L. S., Carlson N. G., Simonato M., Morrow J. D., Roberts J. L., McNamara J. O. Prostaglandin F2alpha is required for NMDA receptor-mediated induction of c-fos mRNA in dentate gyrus neurons. *J. Neurosci.,* 1997; 17: 117-124.

Levy B. D., Kohli P., Gotlinger K., Haworth O., Hong S., Kazani S., Israel., Haley K. J., Serhan C. N. Protectin D1 is generated in asthma and dampens airway inflammation and hyperresponsiveness. *J. Immunol.,* 2007; 178: 496-502.

Li D., Ng A., Mann N. J., Sinclair A. J. Contribution of meat fat to dietary arachidonic acid. *Lipids,* 1998; 33: 437-440.

Li Y., Seifert M. F., Lim S. Y., Salem N. Jr., Watkins B. A. Bone mineral content is positively correlated to *n*-3 fatty acids in the femur of growing rats. *Br. J. Nutr.*, 2010; 104: 674-685.

Libby P. Inflammation and cardiovascular disease mechanisms. *Am. J. Clin. Nutr.,* 2006; 83: 456S-460S.

Lin J., Zhang S. M., Cook N. R., Lee I. M., Buring J. E. Dietary fat and fatty acids and risk of colorectal cancer in women. *Am. J. Epidemiol.,* 2004; 160: 1011-1022.

Lombardo Y. B., Chicco A. G. Effects of dietary polyunsaturated n-3 fatty acids on dyslipidemia and insulin resistance in rodents and humans. A review. *J. Nutr. Biochem.*, 2006; 17: 1-13.

López-Parra M., Titos E., Horrillo R., Ferré N., González-Périz A., Martínez-Clemente M., Planagumà A., Masferrer J., Arroyo V., Clària J. Regulatory effects of arachidonate 5-lipoxygenase on hepatic microsomal TG transfer protein activity and VLDL-triglyceride and apoB secretion in obese mice. *J. Lipid Res.*, 2008; 49: 2513-2523.

Madsen L., Pedersen L. M., Liaset B., Ma T., Petersen R. K., van den Berg S., Pan J., Müller-Decker K., Dülsner E. D., Kleemann R., Kooistra T., Døskeland S. O., Kristiansen K. cAMP-dependent signaling regulates the adipogenic effect of *n*-6 polyunsaturated fatty acids. *J. Biol. Chem.*, 2008; 283: 7196-7205.

Mahmud S., Franco E., Aprikian A. Prostate cancer and use of nonsteroidal anti-inflammatory drugs: systematic review and meta-analysis. *Br. J. Cancer,* 2004; 90: 93-99.

Maillard V., Bougnoux P., Ferrari P., Jourdan M. L., Pinault M., Lavillonnière F., Body G., Le Floch O., Chajès V. N-3 and n-6 fatty acids in breast adipose tissue and relative risk of breast cancer in a case-control study in Tours, France. *Int. J. Cancer*, 2002; 98: 78-83.

Makrides M., Neumann M., Simmer K., Pater J., Gibson R. Are long-chain polyunsaturated fatty acids essential nutrients in infancy? *Lancet,* 1995; 345: 1463-1468.

Mamalakis G., Kafatos A., Kalogeropoulos N., Andrikopoulos N., Daskalopulos G., Kranidis A. Prostate cancer vs hyperplasia: relationships with prostatic and adipose tissue fatty acid composition. *Prostaglandins Leukot. Essent. Fatty Acids,* 2002; 66: 467-477.

Mann N. J., Johnson L. G., Warrick G. E., Sinclair A. J. The arachidonic acid content of the Australian diet is lower than previously estimated. *J. Nutr.,* 1995; 125: 2528-2535.

Männistö S., Pietinen P., Virtanen M. J., Salminen I., Albanes D., Giovannucci E., Virtamo J. Fatty acids and risk of prostate cancer in a nested case-control study in male smokers. *Cancer Epidemiol. Biomarkers Prev.,* 2003; 12: 1422-1428.

Markworth J. F., Cameron-Smith D. Arachidonic acid supplementation enhances in vitro skeletal muscle cell growth via a COX-2-dependent pathway. *Am. J. Physiol. Cell Physiol.*, 2013; 304: C56-C67.

Martinez-Clemente A., Claria J., Titos E. The 5-lipoxygenase/leukotriene pathway in obesity, insulin resistance, and fatty liver disease. *Curr. Opin. Clin. Nutr. Metabol. Care*, 2011; 14: 347-353.

Masters C. L., Simms G., Weinman N. A., Multhaup G., McDonald B. L., Beyreuther K. Amyloid plaque core protein in Alzheimer disease and Down syndrome. *Proc. Natl. Acad. Sci. U S A,* 1985; 82, 4245-4249.

Matz R. L., de Sotomayor M. A., Schott C., Stoclet J. C., Andriantsitohaina R. Vascular bed heterogeneity in age-related endothelial dysfunction with respect to NO and eicosanoids. *Br. J. Pharmacol.*, 2000; 131: 303-311.

McCown T. J., Knapp D. J., Crews F. T. Inferior collicular seizure generalization produces site-selective cortical induction of cyclooxygenase 2 (COX-2). *Brain Res.,* 1997; 767: 370-374.

McDaniel J. C., Belury M., Ahijevych K., Blakely W. Omega-3 fatty acids effect on wound healing. *Wound Repair Regen.,* 2008; 16: 337-345.

McGahon B., Clements M. P., Lynch M. A. The ability of aged rats to sustain long-term potentiation is restored when the age-related decrease in membrane arachidonic acid concentration is reversed. *Neuroscience,* 1997; 81: 9-16.

Miles E. A., Allen E., Calder P. C. In vitro effects of eicosanoids derived from different 20-carbon fatty acids on production of monocyte-derived cytokines in human whole blood cultures. *Cytokine*, 2002; 20: 215-223.

Miyake Y., Sasaki S., Tanaka K., Fukushima W., Kiyohara C., Tsuboi Y., Yamada T., Oeda T., Miki T., Kawamura N., Sakae N., Fukuyama H., Hirota Y., Nagai M; Fukuoka Kinki Parkinson's Disease Study Group. Dietary fat intake and risk of Parkinson's disease: a case-control study in Japan. *J. Neurol. Sci.*, 2010; 288: 117-122.

Mohammed N. A., El-Aleem S. A., El-Hafiz H. A., McMahon R. F. T. Distribution of constitutive (COX-1) and inducible (COX-2) cyclooxygenase in postviral human liver cirrhosis: a possible role for COX-2 in the pathogenesis of liver cirrhosis. *J. Clin. Pathol.*, 2004; 57: 350-354.

Mohrhauer H., Holman R. T. The effect of dose level of essential fatty acids upon fatty acid composition of the rat liver. *J. Lipid Res.,* 1963; 4: 346-350.

Moncada S., Vane J. R. Arachidonic acid metabolites and the interactions between platelets and blood-vessel walls. *N. Engl. J. Med.,* 1979; 300: 1142.

Moon Y. A., Hammer R. E., Horton J. D. Deletion of ELOVL5 leads to fatty liver through activation of SREBP-1c in mice. *J. Lipid Res.,* 2009; 50: 412-423.

Moreno J. J. Differential effects of arachidonic acid and eicosapentaenoic acid-derived eicosanoids on polymorphonuclear transmigration across endothelial cell cultures. *J. Pharmacol. Exp. Ther.*, 2009; 331: 1111-1117.

Morley N., Kuksis A. Postitional specificity of lipoprotein lipase. *J. Biol. Chem.*, 1972; 247: 6389-6393.

Muhlhausler B. S., Ailhaud G. P. Omega-6 polyunsaturated fatty acids and the early origins of obesity. *Curr. Opin. Endocrinol. Diabetes Obes.,* 2013; 20: 56-61.

Muller N., Riedel M., Scheppach C., Brandstatter B., Sokullu S., Krampe K., Ulmschneider M., Engel R. R., Moller H. J., Schwarz M. J. Beneficial antipsychotic effects of celecoxib add-on therapy compared to risperidone alone in schizophrenia. *Am. J. Psychiatry*, 2002; 159: 1029-1034.

Murff H. J., Shu X. O., Li H., Dai Q., Kallianpur A., Yang G., Cai H., Wen W., Gao Y. T., Zheng W. A prospective study of dietary polyunsaturated fatty acids and colorectal cancer risk in Chinese women. *Cancer Epidemiol. Biomarkers Prev.*, 2009; 18: 2283-2291.

Nakano D., Ishii F., Fujii K., Ishikura Y., Akimoto K., Kontani M., Kawashima H., Kiso Y., Matsumura Y. Effects of dietary arachidonic acid supplementation on age-related changes in endothelium-dependent vascular responses. *J. Nutr. Sci. Vitaminol. (Tokyo),* 2007; 53: 75-81.

Nelson G. J., Kelley D. S., Emken E. A., Phinney S. D., Kyle D., Ferretti A. A human dietary arachidonic acid supplementation study conducted in a metabolic research unit: rationale and design. *Lipids,* 1997a; 32: 415-420.

Nelson G. J., Schmidt P. C., Bartolini G., Kelley D. S., Kyle D. The effect of dietary arachidonic acid on platelet function, platelet fatty acid composition, and blood coagulation in humans. *Lipids,* 1997b; 32: 421-425.

Nelson G. J., Schmidt P. C., Bartolini G., Kelley D. S., Phinney S. D., Kyle D., Silbermann S., Schaefer E. J. The effect of dietary arachidonic acid on plasma lipoprotein distributions, apoproteins, blood lipid levels, and tissue fatty acid composition in humans. *Lipids,* 1997c; 32: 427-433.

Nery F. G., Monkul E. S., Hatch J. P., Fonseca M., Zunta-Soares G. B., Frey B. N., Bowden C. L., Soares J. C. Celecoxib as an adjunct in the treatment of depressive or mixed episodes of bipolar disorder: a double-blind, randomized, placebo-controlled study. *Hum. Psychopharmacol.*, 2008; 23: 87-94.

NHANES. The National Health and Nutrition Examination Survey 2005-2006. Internet source. Available from: http://appliedresearch.cancer/diet/foodsources/fatty_acids/table4.html (accessed 06.01.14).

Nielsen M. S., Gammelmark A., Madsen T., Obel T., Aardestrup I., Schmidt E. B. The effect of low-dose marine *n*-3 fatty acids on the biosynthesis of pro-inflammatory 5-lipoxygenase pathway metabolites in overweight subjects: a randomized controlled trial. *Prostaglandins Leukot. Essent. Fatty Acids,* 2012; 87: 43-48.

Nigam S., Fiore S., Luscinskas F. W., Serhan C. N. Lipoxin A4 and lipoxin B4 stimulate the release but not the oxygenation of arachidonic acid in human neutrophils: dissociation between lipid remodeling and adhesion. *J. Cellular Physiol.,* 1990; 143: 512-523.

Nkondjock A., Shatenstein B., Maisonneuve P., Ghadirian P. Assessment of risk associated with specific fatty acids and colorectal cancer among French-Canadians in Montreal: a case-control study. *Int. J. Epidemiol.,* 2003a; 32: 200-209.

Nkondjock A., Shatenstein B., Ghadirian P. A case-control study of breast cancer and dietary intake of individual fatty acids and antioxidants in Montreal, Canada. *Breast,* 2003b; 12: 131-138.

Node K., Huo Y., Ruan X., Yang B., Spiecker M., Ley K., Zeldin D. C., Liao J. K. Anti-inflammatory properties of cytochrome P450 epoxygenase-derived eicosanoids. *Science,* 1999; 285: 1276-1279.

Oe H., Hozumi T., Murata E., Matsuura H., Negishi K., Matsumura Y., Iwata S., Ogawa K., Sugioka K., Takemoto Y., Shimada K., Yoshiyama M., Ishikura Y., Kiso Y., Yoshikawa J. Arachidonic acid and docosahexaenoic acid supplementation increases coronary flow velocity reserve in Japanese elderly individuals. *Heart,* 2008; 94: 316-321.

Ojeda S. R., Urbanski H. F., Junier M. P., Capdevila J. The role of arachidonic acid and its metabolites in the release of neuropeptides. *Ann. N Y Acad. Sci.,* 1989; 559: 192-207.

Oltman C. L., Weintraub N. L., VanRollins M., Dellsperger K. C. Epoxyeicosatrienoic acids and dihydroxyeicosatrienoic acids are potent vasodilators in the canine coronary microcirculation. *Circ. Res.,* 1998; 83: 932-939.

Ott S. M. Diet for the heart or the bone: a biological tradeoff. *Am. J. Clin. Nutr.,* 2004; 79: 4-5.

Patterson E., Wall R., Fitzgerald G. F., Ross R. P., Stanton C. Health implications of high dietary omega-6 polyunsaturated fatty acids. *J. Nutr. Metab.,* 2012; 2012: 539-426.

Pérez-Chacón G., Astudillo A. M., Balgoma D., Balboa M. A., Balsinde J. Control of free arachidonic acid levels by phospholipases A2 and lysophospholipid acyltransferases. *Biochim. Biophys. Acta,* 2009; 1791: 1103-1113.

Petrik M. B., McEntee M. F., Chiu C. H., Whelan J. Antagonism of arachidonic acid is linked to the antitumorigenic effect of dietary eicosapentaenoic acid in ApcMin/1 mice. *J. Nutr.* 2000; 130: 1153-1158.

Pettipher R, Hansel TT. Antagonists of the prostaglandin D2 receptor CRTH2. *Drug News Perspect.,* 2008; 21: 317-322.

Phillis J. W., O'Regan M. H. The role of phospholipases, cyclooxygenases, and lipoxygenases in cerebral ischemic/traumatic injuries. *Crit. Rev. Neurobiol.,* 2003; 15: 61-90.

Phillis J. W., Horrocks L. A., Farooqui A. A. Cyclooxygenases, lipoxygenases, and epoxygenases in CNS: Their role and involvement in neurological disorders. *Brain Res.,* 2006; 52: 201-243.

Pickard J. D. Role of prostaglandins and arachidonic acid derivatives in the coupling of cerebral blood flow to cerebral metabolism. *J. Cereb. Blood Flow Metab.,* 1981; 1: 361-384.

Pulichino A. M., Rowland S., Wu T., Clark P., Xu D., Mathieu M. C., Riendeau D., Audoly L. P. Prostacyclin antagonism reduces pain and inflammation in rodent models of hyperalgesia and chronic arthritis. *J. Pharmacol. Exp. Ther.,* 2006; 319: 1043-1050.

Rapoort S. Arachidonic acid and the brain. *J. Nutr.,* 2008; 138: 2515-2520.

Ricciotti E., Fitzgerald G. A. Prostaglandins and inflammation. *Arterioscler. Thromb. Vasc. Biol.,* 2011; 31: 986-1000.

Roberts M. D., Iosia M., Kerksick C. M., Taylor L. W., Campbell B., Wilborn C. D., Harvey T., Cooke M., Rasmussen C., Greenwood M., Wilson R., Jitomir J., Willoughby D., Kreider R. B. Effects of arachidonic acid supplementation on training adaptations in resistance-trained males. *J. Int. Soc. Sports Nutr.,* 2007; 4: 21.

Roman R. J. P-450 metabolites of arachidonic acid in the control of cardiovascular function. *Physiol. Rev.,* 2002; 82: 131-185.

Russo G. L. Dietary *n*-6 and *n*-3 polyunsaturated fatty acids: from biochemistry to clinical implications in cardiovascular prevention. *Biochem. Pharmacol.,* 2009; 77: 937-946.

Saadatian-Elahi M., Toniolo P., Ferrari P., Goudable J., Akhmedkhanov A., Zeleniuch-Jacquotte A., Riboli E. Serum fatty acids and risk of breast cancer in a nested case-control study of the New York University

Women's Health Study. *Cancer Epidemiol. Biomarkers Prev.,* 2002; 11: 1353-1360.

Sakai M., Kakutani S., Horikawa C., Tokuda H., Kawashima H., Shibata H., Okubo H., Sasaki S. Arachidonic acid and cancer risk: a systematic review of observational studies. *BMC Cancer,* 2012; 12: 606.

Sanak M., Levy B. D., Clish C. B., Chiang N., Gronert K., Mastalerz L., Serhan C. N., Szczeklik A. Aspirin-tolerant asthmatics generate more lipoxins than aspirin-intolerant asthmatics. *Eur. Respir. J.,* 2000; 16: 44-49.

Sanchez-Mejia R. O., Newman J. W., Toh S., Yu G. Q., Zhou Y., Halabisky B., Cissé M., Scearce-Levie K., Cheng I. H., Gan L., Palop J. J., Bonventre JV, Mucke L. Phospholipase A2 reduction ameliorates cognitive deficits in a mouse model of Alzheimer's disease. *Nat. Neurosci., 2008;* 11: 1311-1318.

Sanders T. A. B., Roshanai F. Platelet phospholipid fatty acid composition in vegans compared with age- and sex-matched omnivore controls. *Aur. J. Clin. Nutr.,* 1992; 46: 823-831.

Sapirstein A., Bonventre J. V. Phospholipases A2 in ischemic and toxic brain injury. *Neurochem. Res.,* 2000; 25: 745-753.

Schaeffer E. L., Forlenza O. V., Gattaz W. F. Phospholipase A2 activation as a therapeutic approach for cognitive enhancement in early-stage Alzheimer disease. *Psychopharmacology,* 2009; 202: 37-51.

Schmitz G., Ecker J. The opposing effects of *n*-3 and *n*-6 fatty acids. *Prog. Lipid Res.,* 2008; 47: 147-155.

Sellers M. M., Stallone J. N. Sympathy for the devil: the role of thromboxane in the regulation of vascular tone and blood pressure. *Am. J. Physiol. Heart Circ. Physiol.,* 2008; 294: H1978-H1986.

Serhan C. N., Hamberg M., Samuelsson B. Lipoxins: novel series of biologically active compounds formed from arachidonic acid in human leukocytes. *Proc. Natl. Acad. Sci. U S A,* 1984; 81: 5335-5339.

Serhan C. N. Lipoxins and novel aspirin-triggered 15-epi-lipoxins (ATL): a jungle of cell-cell interactions or a therapeutic opportunity? *Prostaglandins,* 1997; 53: 107-137.

Serhan C. N., Jain A., Marleau S., Clish C., Kantarci A., Behbehani B., Colgan S. P., Stahl G. L., Merched A., Petasis N. A., Chan L., Van Dyke T. E. Reduced inflammation and tissue damage in transgenic rabbits overexpressing 15-lipoxygenase and endogenous anti-inflammatory lipid mediators. *J. Immunol.,* 2003; 171: 6856-6865.

Serhan C. N., Chiang N. Novel endogenous small molecules as the checkpoint controllers in inflammation and resolution: entree for resoleomics. *Rheum. Dis. Clin. North Am.,* 2004; 30: 69-95.

Serhan C. N., Savill J. Resolution of inflammation: the beginning programs the end. *Nat. Immunol*., 2005; 6: 1191-1197.

Serhan C. N. Lipoxins and aspirin-triggered 15-epi-lipoxins are the first lipid mediators of endogenous anti-inflammation and resolution. *Prostaglandins Leukot. Essent. Fatty Acids,* 2005; 73: 141-162.

Serhan C. N., Yang R., Martinod K., Kasuga K., Pillai P. S., Porter T. F., Oh S. F., Spite M. Maresins: novel macrophage mediators with potent anti-inflammatory and proresolving actions. *J. Exp. Med.,* 2009; 206: 15-23.

Seyberth H. J., Oelz O., Kennedy T., Sweetman B. J., Danon A., Frich J. C., Heimberg M., Oates J. A. Increased arachidonate in lipids after administration to man: effects on prostaglandin biosynthesis. *Clin. Pharmacol. Ther.,* 1975; 8: 519-521.

Shen C. L., Peterson J., Tatum O. L., Dunn D. M. Effect of long-chain n-3 polyunsaturated fatty acid on inflammation mediators during osteoblastogenesis. *J. Med. Food*, 2008; 11: 105-110.

Simopoulos A. P. Evolutionary aspects of diet and essential fatty acids. *World Rev. Nutr. Diet,* 2001; 88: 18-27.

Simopoulos A. P. Evolutionary aspects of diet, the omega-6/omega-3 ratio and genetic variation: nutritional implications for chronic diseases. *Biomed. Pharmacother.,* 2006; 60: 502-507.

Sinclair A. J., O'Dea K. The significance of arachidonic acid in hunter-gatherer diets: implications for the contemporary Western diet. *J. Food Lipids,* 1993; 1: 142-157.

Sioen I. A., Pynaert I., Matthys C., De Backer G., Van Camp J., De Henauw S. Dietary intakes and food sources of fatty acids for Belgian women, focused on *n*-6 and *n*-3 polyunsaturated fatty acids. *Lipids,* 2006; 41: 415-422.

Smit E. N., Koopmann M., Boersma E. R., Muskiet F. A. Effect of supplementation of arachidonic acid (AA) or a combination of AA plus docosahexaenoic acid on breast milk fatty acid composition. *Prostaglandins Leukot. Essent. Fatty Acids*, 2000; 62: 335-40.

Spector A. A. Plasma free fatty acid and lipoproteins as sources of polyunsaturated fatty acid for the brain. *J. Mol. Neurosci.,* 2000; 16: 159-165.

Stables M. J., Gilroy D. W. Old and new generation lipid mediators in acute inflammation and resolution. *Prog. Lipid Res.,* 2011; 50: 35-51.

Stoffel W., Holz B., Jenke B., Binczek E., Günter R. H., Kiss C., Karakesisoglou I., Thevis M., Weber A-A., Arnhold S., Addicks K. Δ6-Desaturase (FADS2) deficiency unveils the role of ω3- and ω6-polyunsaturated fatty acids. *EMBO J.*, 2008; 27: 2281-2292.

Sublette M. E., Russ M. J., Smith G. S. Evidence for a role of the arachidonic acid cascade in affective disorders: a review. *Bipolar Disord.*, 2004; 6: 95-105.

Taber L., Chiu C. H., Whelan J. Assessment of the arachidonic acid content in foods commonly consumed in the American diet. *Lipids,* 1998; 33: 1151-1157.

Takkouche B., Regueira-Méndez C., Etminan M. Breast cancer and use of nonsteroidal anti-inflammatory drugs: a meta-analysis. *J. Natl. Cancer Inst.*, 2008; 100: 1439-1447.

Teismann P., Tieu K., Choi D. K., Wu D. C., Naini A., Hunot S., Vila M., Jackson-Lewis V., Przedborski S. Cyclooxygenase-2 is instrumental in Parkinson's disease neurodegeneration. *Proc. Natl. Acad. Sci. U S A,* 2003; 100: 5473-5478.

Thies F., Miles E. A., Nebe-von-Caron G., Powell J. R., Hurst T. L., Newsholme E. A., Calder P. C. Influence of dietary supplementation with long-chain *n*-3 or *n*-6 polyunsaturated fatty acids on blood inflammatory cell populations and functions and on plasma soluble adhesion molecules in healthy adults. *Lipids,* 2001a; 36: 1183-1193.

Thies F., Nebe-von-Caron G., Powell J. R., Yaqoob P., Newsholme E. A., Calder P. C. Dietary supplementation with γ-linolenic acid or fish oil decreases T lymphocyte proliferation in healthy older humans. *J. Nutr.*, 2001b; 131: 1918-1927.

Thusgaard M., Christensen J. H., Mørn B., Andersen T. S., Vige R., Arildsen H., Schmidt E. B., Nielsen H. Effect of fish-oil (*n*-3 polyunsaturated fatty acids) on plasma lipids, lipoproteins and inflammatory markers in HIV-infected patients treated with antiretroviral therapy: a randomized, double-blind, placebo-controlled study. *Scand. J. Infect. Dis.*, 2009; 41: 760-766.

Tilley S. L., Coffman T. M., Koller B. H. Mixed messages: modulation of inflammation and immune responses by prostaglandins and thromboxanes. *J. Clin. Invest.*, 2001; 108: 15-23.

Tjonahen E., Oh S. F., Siegelman J., Elangovan S., Percarpio K. B., Hong S., Arita M., Serhan C. N. Resolvin E2: identification and anti-inflammatory actions: pivotal role of human 5-lipoxygenase in resolvin E series biosynthesis. *Chem. Biol.*, 2006; 13: 1193-1202.

Townsend K. P., Praticò D. Novel therapeutic opportunities for Alzheimer's disease: focus on nonsteroidal anti-inflammatory drugs. *FASEB J.,* 2005; 19: 1592-1601.

Trappe T. A., Fluckey J. D., White F., Lambert C. P., Evans W. J. Skeletal muscle PGF(2)(alpha) and PGE(2) in response to eccentric resistance exercise: influence of ibuprofen acetaminophen. *J. Clin. Endocrinol. Metab.,* 2001; 86: 5067-5070.

Ukoli F. A., Fowke J. H., Akumabor P., Oguike T., Taher K. A., Murff H. J., Amaefuna E. R., Kittles R., Ahaghotu C., Osime U., Beech D. J. The association of plasma fatty acids with prostate cancer risk in African Americans and Africans. *J. Health Care Poor Underserved,* 2010; 21: 127-147.

Vachier I., Chanez P., Bonnans C., Godard P., Bousquet J., Chavis C. Endogenous anti-inflammatory mediators from arachidonate in human neutrophils. *Biochem. Biophys. Res. Commun.*, 2002; 290: 219-224.

van Goor S. A., Dijck-Brouwer D. A., Hadders-Algra M., Doornbos B., Erwich J. J., Schaafsma A., Muskiet F. A. Human milk arachidonic acid and docosahexaenoic acid contents increase following supplementation during pregnancy and lactation. *Prostaglandins Leukot. Essent. Fatty Acids,* 2009; 80: 65-69.

Vandenburgh H. H., Shansky J., Karlisch P., Solerssi R. L. Mechanical stimulation of skeletal muscle generates lipid-related second messengers by phospholipase activation. *J. Cell Physiol.*, 1993; 155: 63-71.

Videla, L. A., Rodrigo R., Araya J., Poniachik J. Oxidative stress and depletion of hepatic long-chain polyunsaturated fatty acids may contribute to nonalcoholic fatty liver disease. *Free Radic. Biol. Med.,* 2004; 37: 1499-1507.

Videla L. A., Rodrigo R., Araya J., Poniachik J. Insulin resistance and oxidative stress interdependency in non-alcoholic fatty liver disease. *Trends Mol. Med.*, 2006; 12: 555-558.

Voorrips L. E., Brants H. A., Kardinaal A. F., Hiddink G. J., van den Brandt P. A., Goldbohm R. A. Intake of conjugated linoleic acid, fat, and other fatty acids in relation to postmenopausal breast cancer: the Netherlands Cohort Study on Diet and Cancer. *Am. J. Clin. Nutr.,* 2002; 76: 873-882.

Vuppalanchi R., Cummings O. W., Saxena R., Ulbright T. M., Martis N., Jones D. R., Bansal N., Chalasani N. Relationship among histologic, radiologic, and biochemical assessments of hepatic steatosis: a study of human liver samples. *J. Clin. Gastroenterol.*, 2007; 41: 206-210.

Wainwright P. E. Dietary essential fatty acids and brain function: a developmental perspective on mechanisms. *Proc. Nutr. Soc.*, 2002, 61: 61-69.

Wall R., Ross R. P., Fitzgerald G. F., Stanton C. Fatty acids from fish: the anti-inflammatory potential of long-chain omega-3 fatty acids. *Nutr. Rev.*, 2010; 68: 280-289.

Wang M. H., Guan H., Nguyen X., Zand B. A., Nasjletti A., Laniado-Schwartzman M. Contribution of cytochrome P-450 4A1 and 4A2 to vascular 20-hydroxyeicosatetraenoic acid synthesis in rat kidneys. *Am. J. Physiol.,* 1999; 276: F246-F253.

Wang Z. J., Liang C. L., Li G. M., Yu C. Y., Yin M. Neuroprotective effects of arachidonic acid against oxidative stress on rat hippocampal slices. *Chem. Biol. Interact.*, 2006; 163: 207-217.

Wang D., DuBois R. N. Eicosanoids and cancer. *Nat. Rev. Cancer*, 2010; 10: 181-193.

Weiler H. A. Polyunsaturated fatty acids and bone mass. *Am. J. Clin. Nutr.,* 2004; 80: 1085-1086.

Werz O., Steinhilber D. Development of 5-lipoxygenase inhibitors – lessons from cellular enzyme regulation. *Biochem. Pharmacol.,* 2005; 70: 327-333.

Weseler A. R., Dirix C. E. H., Bruins M. J., Hornstra G. Dietary arachidonic acid dose-dependently increases the arachidonic acid concentration in human milk. *J. Nutr.*, 2008; 138: 2190-2197.

Whelan J., McEntee M. F. Dietary (*n*-6) PUFA and intestinal tumorigenesis. *J. Nutr.,* 2004; 134: 3421S-3426S.

Williams J. H., Errington M. L., Lynch M. A., Bliss T. V. P. Arachidonic acid induces a long-term activity-dependent enhancement of synaptic transmission in the hippocampus. *Nature*, 1989; 341: 739-742.

In: Arachidonic Acid
Editor: Jason M. O'Keefe
ISBN: 978-1-63117-619-7

Chapter 3

CONTROL OF INTRACELLULAR LEVELS OF FREE ARACHIDONIC ACID: A TARGET IN THERAPEUTIC OR PREVENTIVE STRATEGIES AGAINST ALZHEIMER'S DISEASE?

Mélanie H. Thomas, Sandra Pelleieux, Ahmad Allouche, Julie Colin, Thierry Oster, Catherine Malaplate-Armand and Jean Luc Olivier
URAFPA-BFLA, INRA USC 0340, Université de Lorraine,
Nancy, France

ABSTRACT

Arachidonic acid (ARA) is one of the two most abundant polyunsaturated fatty acids (PUFAs) in brain, especially in neuronal cells, and can be a target in the fight against the major public health concern represented by Alzheimer's disease (AD) and related disorders in high- as well as in low-income countries where elders constitute a growing part of the population. The numerous studies devoted to the AD pathophysiological mechanisms emphasized the role of the amyloid-β (Aβ) oligomers generated from the precursor protein APP. Aβ oligomer-induced disruption of hippocampal synapses is recognized as the first cellular alteration in AD, leading to reduced cognitive abilities. Furthermore, the occurrence of neuroinflammation and oxidative stress is strongly linked to the AD process. But the mechanisms responsible for

AD initiation are still largely unknown, although their identification is critical to design efficient therapeutic or preventive strategies. ARA was first considered in AD as a mediator of neuroinflammation and oxidative stress through its conversion into pro-inflammatory eicosanoids by several enzymes such as 5-lipoxygenase and cyclooxygenase-2. Upregulation of these enzymes in AD brain could contribute to the pathophysiological process by amplifying neuroinflammation and by increasing Aβ oligomer production and neurotoxicity. Other studies showed that free ARA is involved in essential synaptic functions through its effects on several protein kinases or its interactions with SNARE components in the traffic of synaptic vesicles. Therefore, free intracellular ARA levels must be adjusted to physiological needs at adequate cellular locations. ARA is specifically released from membrane phospholipids by cytosolic phospholipase A_2 ($cPLA_2$-α). We and other groups demonstrated that $cPLA_2$-α is an upstream effector of the Aβ oligomer-induced apoptotic cascade, while neuroprotective effects are observed in $cPLA_2$-α KO models or in response to pharmacological inhibitors. Increase in intraneuronal free ARA levels could also result from the catabolism of endocannabinoids by monoacylglycerol lipase (MAGL). The modification of the activities of MAGL and other endocannabinoid-hydrolyzing enzymes in AD could reduce the putative positive effects of the endocannabinoid signaling. Interestingly, several acyl-CoA synthetases or lysophospholipid acyltransferases were recently characterized and could antagonize the effects of $cPLA_2$-α and MAGL on the free ARA levels in brain cells. Furthermore, ARA is believed to compete with incorporation of protective docosahexaenoic acid (DHA) into the membrane phospholipids. The unbalance between omega-3 and omega-6 PUFAs in Western diets potentially favors the increase in brain ARA pool compared to that of DHA. All these findings suggest that controlling the levels of intracellular free ARA could be of valuable interest in therapeutic or preventive strategies against AD, as we discussed in this chapter.

INTRODUCTION

Alzheimer's disease (AD) is characterized by a progressive loss of memory associated with various behavior perturbation and is the main form of senile dementia (about 60% of the cases). Apart some rare familial forms (less than 1% of the cases), the large majority of the AD forms are sporadic. The causes of these sporadic forms are still unknown and there is no clear risk factor except aging. Therefore, AD and other senile dementia constitute a major health concern not only in developed countries but also in all the low-

and mid-income level countries in which elders represent a growing part of their population. In 2010, 35.6 millions of people were affected by senile dementia and about 50% of these patients live in high-income countries. The number of affected people will exceed 65 million in 2030, with about 60% in low- and mid-income levels (International World Alzheimer report 2010). The total estimated worldwide costs of dementia are US$604 billion in 2010, i.e., around 1% of the world's gross domestic product. In front of these dramatic epidemiological predictions, there is no efficient therapeutic treatment or established preventive strategy. The fight against AD requires a better understanding of its mechanisms which are still largely unknown. Studies about AD mechanisms were historically based on the initial observation of two pathognomonic histological signs in AD patient brain and the identification of their main constituents: neurofibrillar tangles formed by the aggregation of hyperphosphorylated Tau protein and amyloid plaques resulting from the deposit of amyloid-β (Aβ) peptide. These two signs result from complex mechanisms including deregulation of protein kinases, alteration of protein maturation and transportation, synapse dysfunctions and neuroinflammation. The roles of lipids and peculiarly arachidonic acid (ARA) and its derivatives were originally deduced from the finding that neuroinflammation plays a major role in AD process. But ARA and some of its derivatives such as endocannabinoids interfere with synapse functions which are altered in the early steps of AD. Furthermore, AD also seems to modify ARA homeostasis in membrane, which can lead to enhancement of Aβ peptide production and neurotoxicity. In this chapter, we will first examine the main knowledge about the AD mechanisms by focusing on synaptotoxicity and neuroinflammation, their relationship with ARA derivatives and the various enzymes generating them. We will then study how ARA itself is involved in synaptic functions and whether the balance between its release and its incorporation into membranes could be a target in therapeutic or preventive strategies against AD.

Alzheimer's Disease Mechanisms: A Lipid and Membrane Story

The Aβ peptide forming the amyloid plaques in hippocampus and cortex of AD patients is generated by the sequential cleavage of the amyloid precursor protein (APP) by two enzymes, the β- and γ-secretases (Figure 1). APP is a transmembrane protein located in synaptic membrane of neuronal cells. Its gene is ubiquitously expressed and alternative splicing generates several forms in peripheral cells, although only a specific 695-aminoacid form is expressed in neurons. Its neuronal functions are still unclear, but there are some evidences that APP and its paralogs APLP1/2 can be involved in neuronal survival and synaptic communication (Dawson *et al.*, 1999), thereby playing an important role in synaptic plasticity and supporting memory formation and consolidation (Korte *et al.*, 2012). Alternatively, APP is hydrolyzed by α- and γ-secretases which generates the shorter and non-neurotoxic Aα-peptide (Figure 1). In physiological conditions, the hydrolysis of APP by α/γ-secretases activities would represent the main pathway compared to the amyloidogenic pathway involving β/γ-secretases (Thinakaran & Koo, 2008). Two β-secretases, BACE-1 and BACE-2 have been cloned (Dingwall, 2001), while γ-secretase activity is supported by a 3-protein complex, presenilin-1 and -2 [PS1/2], nicastrin, PEN or APH-1 (Kimberly et *al.*, 2003). Such a complex might be required to cleave APP in the hydrophobic core of the membrane, which is quite exceptional in protease activities. In addition, γ-secretase can cleave APP at two sites, subsequently releasing two distinct peptides long of 40 or 42 amino acids [$A\beta_{1-40}$ or $A\beta_{1-42}$], the first one being the major but less neurotoxic form (Thinakaran & Koo, 2008). Other minor truncated or modified Aβ peptide species are also produced in brain, but their roles in brain physiology and AD are still poorly identified. γ-secretase have other substrates than APP such as Notch (Vetrivel *et al.*, 2006). Besides β- and γ-secretases, 3 members of the ADAM metalloproteinase family have been characterized as having α-secretase activities (Vingtdeux & Marambaud, 2012). These various secretases also generate N- and C-terminal products. The N-terminal soluble product sAPPα displays neurotrophic activities (Furakawa *et al.*, 1994), while the sAPPβ peptide generated can participate to neuronal death as ligand of death receptors (Nikolaev *et al.*, 2009). This difference of biological activities is striking since these two peptides differ only by a 16 amino-acid sequence present at the C-terminal end of the sAPPα peptide. In addition, the intracellular C-terminal fragment of APP released by the γ-secretase complex can be translocated into the nucleus to act as a transcription factor (Chang & Suh, 2010). The rare AD familial forms are linked to APP gene duplication or to mutations in APP or PS genes, which induces large overproduction of Aβ peptide prone to favor Aβ oligomerization/aggregation into toxic assemblies. But whether these mutations induce gain or loss of functions is still a debate. At least some APP gene mutations modify its subcellular location, promoting its catabolism by β- and γ-secretases. Similarly, in sporadic AD, alterations of APP location and trafficking might modify the Aβ peptide production. The non-amyloidogenic pathway of APP hydrolysis would occur at the synaptic membrane, whereas APP would interact with β- and γ-secretases in some endosome species during its recycling between the synaptic membrane and endoplasmic reticulum, thus generating Aβ peptide (Tang 2009). The importance of this recycling process in the production of Aβ peptide from APP has been emphasized by the identification of the role of SORL1 and SORLA proteins, two members of the lipoprotein receptor superfamily, in the endosomal trafficking and by the correlation of their mutations to AD occurrence risk in some populations (Rogaeva *et al.*, 2007; Willnow *et al.*, 2010; Reitz *et al.*, 2011).

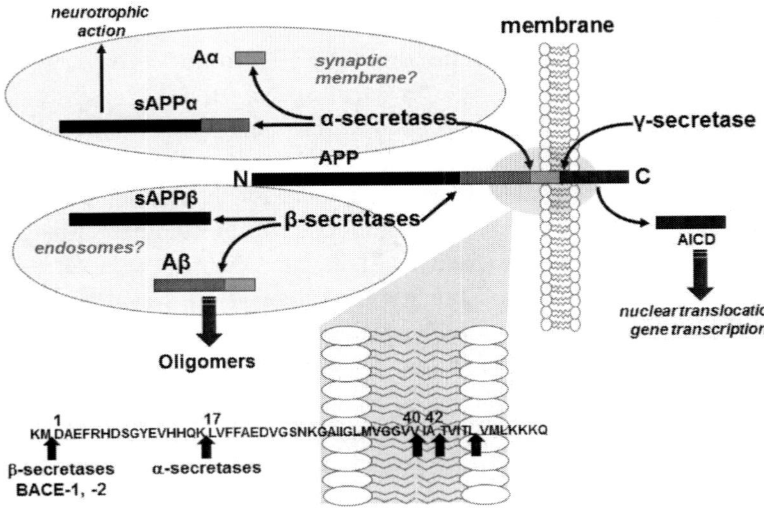

Figure 1. Hydrolysis of APP by the α-, β- and γ- secretases and the resulting peptides. Non-amyloidogenic and amyloidogenic pathways.

For a long time, most of the works about AD were focused on Aβ peptide production and its aggregation into amyloid plaques. Therefore, most of the therapeutic strategies were developed with the aim to inhibit plaque formation or to favor their disaggregation. Microglial cells and astrocytes are attracted and proliferate around the mature amyloid plaques. This neuroinflammatory reaction involves the synthesis of lipid mediators and the production of reactive oxygen species (ROS) which are assumed to induce the death of surrounding neurons. But several experimental data disfavor an exclusive role for the amyloid plaques, at least in the early steps of the disease. For example, some murine AD models expressing the mutated human APP gene display cognitive and synaptic cellular alterations a long time before the occurrence of the first amyloid plaques (Hsia *et al.*, 1999; Mucke *et al.*, 2000; Wu *et al.*, 2004). In humans, the number of plaques and their content in Aβ peptide do not correlate with the severity of the symptoms, which on the contrary depends on the brain concentrations of soluble Aβ oligomers (Mc Lean *et al.*, 1999). It is now well admitted that these oligomers are the main agents of AD, especially in early AD (Drouet *et al.*, 2000; Walsh & Selkoe, 2004). Aβ peptide forms various oligomer species, from dimers, trimers to higher molecular range species such as dodecamers (Larson & Lesné, 2012) or Aβ-derived diffusible ligands (ADDLs) (Krafft & Klein, 2010). Several oligomeric species are formed in human brain during aging, target synaptic membranes in which they colocalize with the PSD95 complex (Lacor *et al.*, 2004) and induce the first synaptic dysfunctions (Rowan *et al.*, 2007). Some authors postulate that dimers are the main agents of AD (Jin *et al.*, 2011), while others assume that several species contribute to the whole neurotoxic effect in AD (Klyubin *et al.*, 2012). Although it is clear that interactions of Aβ oligomers with neuronal membranes are critical for their cellular effects, two hypotheses are still being debated for the mediation of the neurotoxic effects of Aβ oligomers: either the existence of putative membrane receptors, either a direct modification of the membrane organization itself. Regarding the latter hypothesis, Aβ peptide oligomers can induce the fusion of membranes (Pillot *et al.*, 1996) and this fusogenic ability is based on the peculiar conformation of the peptide in solution (Morgan *et al.*, 2004). In addition, Aβ peptide oligomers are able to form channels in the membranes of several cell types including neurons (Fraser *et al.*, 1997). These channels allow calcium influx which initiates several pro-apoptotic pathways (Kagan *et al.*, 2002). Several putative receptors have also been described for the Aβ oligomers, including the prion protein [PrPc] (Laurén *et al.*, 2009) and the APP protein itself or its paralogs (Shaked *et al.*, 2006). The fact that Aβ oligomers represent very complex mixtures in solution does not facilitate the confirmation that these putative receptors effectively mediate the Aβ peptide effects. Finally, it is worthy to remind that Aβ peptide is generated within the neuronal cells and an endoplasmic reticulum stress or mitochondrial dysfunctions are induced by these intraneuronal Aβ oligomers (Umeda *et al.*, 2011). After their production, Aβ peptide is degraded in the neuronal lysosomes or extracellularly by neprilysine or insuline degrading enzyme [IDE] (Wang *et al.*, 2006). Aβ peptides are also exported to blood through the blood-brain barrier (BBB) to avoid its brain accumulation. Several works showed the involvement of lipoprotein receptor family members such as LRP1 in this exportation processes (Sagare *et al.*, 2007).

Neurofibrillary tangles are correlated with the severity of AD and their number and extent in the brain are the basis of the Braak classification of AD evolution (Serrano-Pozo *et al.*, 2011). They are formed inside the neuronal cells by aggregation of the various Tau isoforms. These Tau isoforms are members of the microtubule-associated proteins (MAP) which constitute the neuronal cytoskeleton. Hyperphosphorylation of Tau on several non-physiological sites modifies its conformation and causes its aggregation in neurofibrillary tangles (Köpke *et al.*, 1993). The resulting disorganization of the axonal transportations from the nucleus to the synapse leads to the neuronal degeneration and death. Tauopathies are present in various neurodegenerative diseases and senile dementia (Brandt *et al.*, 2005). The signaling pathways leading to Tau hyperphosphorylation in AD are still elusive and should be the result of several dysregulations. Aβ oligomers affect Tau phosphorylation through several pathways. For example, the PrPc protein binds to Aβ oligomers and Fyn at the dendritic spines and the formation of this complex activates the Tau phosphorylation by Fyn (Wang *et al.*, 2013a). The partitioning defective-1 (PAR-1) and the microtubule affinity-regulating kinases (MARK) have also been identified as mediating the effect of Aβ oligomers on Tau phosphorylation and toxicity on synapses (Yu *et al.*, 2012). The mice expressing the E963Δ APP mutant accumulate low molecular-weight Aβ oligomers and display hyperphosphorylated Tau isoforms inside the neuronal cells at the age of 8 months (Tomiyama *et al.*, 2010).

As shown above, Aβ oligomers are critical agents of AD whose production and neurotoxic effects mostly rely on membrane constituents and functions. Furthermore, it is important to notice that AD progresses over several years or decades. One or several yet unidentified events lead to Aβ accumulation in the brain through its overproduction or decrease or its clearance. This results in the early synaptic dysfunctions and then in the initiation of a vicious cycle involving Aβ oligomer accumulation, neuroinflammation, vascular alterations and neuronal cell death (Figure 2). Metabolic diseases such as obesity and diabetes increase the risk of occurrence of AD (Carlsson, 2010; Luchsinger *et al.*, 2012). These diseases involve dysregulation of several hormones, insulin, ghrelin and leptin, which also contribute to the regulation of Aβ production. Metabolic diseases also involve low-levels of general inflammation with circulating inflammation markers (Choi *et al.*, 2013). It has been shown that the induction of peripheral inflammation can stimulate neuroinflammation in brain which could contribute to the initiation of neurodegenerative diseases such as AD. Prostaglandins and leukotrienes are produced by ARA conversion and are powerful mediators of inflammation. Furthermore, the endocannabinoids, which are also ARA derivatives, are synaptic retro-messengers. How these various molecules participate to the AD process is described below.

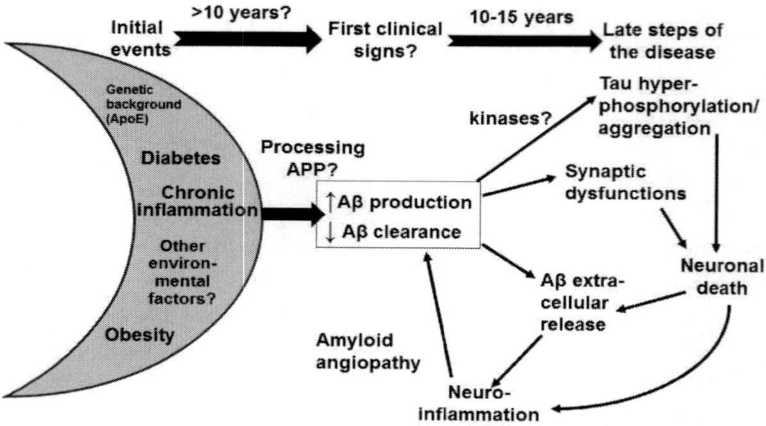

Figure 2. Risk factors and interplay of the various pathological events in AD.

ROLE OF ARACHIDONIC ACID DERIVATIVES IN ALZHEIMER'S DISEASE: LEUKOTRIENES, PROSTAGLANDINS AND ENDOCANNABINOIDS

ARA is converted by the cyclooxygenase-1 (COX-1) and -2 (COX-2) enzymes into prostaglandin PGH2 which is the substrate of various prostaglandin synthases to give the diverse prostaglandins. The involvement of this pathway was first suggested by the fact that long-term treatment by nonsteroidal anti-inflammatory drugs (NSAIDs) decreases the AD incidence (In't Veld *et al.*, 1998). Since NSAIDs are potent inhibitors of COX-1 and -2, many studies were devoted to the expression and activity of these enzymes. COX-2 is usually seen as an inducible enzyme whose expression is increased in inflammation in most of the tissues and organs. However, brain is exceptional in the sense that COX-2 mRNA and protein are constitutively expressed at least in some neurons and during development (Malinska *et al.*, 1999; Minghetti, 2004), which indicates that this enzyme is required for basic functions such as synaptogenesis and memory consolidation. Lukiw & Bazan (1997) first observed a trend for higher COX-2 mRNA levels in the neocortex of 10 AD patients compared to 15 control subjects, but noticed in both groups large individual variations that they associated with differences in inflammatory processes. Thereafter, Yasojima et al., (1999) described an overexpression in pyramidal neurons of AD patients. Ho et al., (2001) and Fujimi et al., (2007) studied immunoreactivity of COX-2 in hippocampus of several tens of AD patients and control subjects and confirmed a COX-2 overexpression in pyramidal neurons of AD patients in this brain region. Both research groups found a correlation between the severity of the disease and the level of COX-2 overexpression in CA1 to CA3 subdivisions of hippocampus, but differed on the precise subdivision in which this correlation was established, CA1 (Fujimi et al., 2007) or CA2/CA3 (Ho et al., 2001). COX-2 overexpression in neuronal cells suggests that this enzyme is involved in neuronal AD-dependent mechanisms rather than in glial cell-dependent inflammatory mechanisms as initially assumed. Indeed, $A\beta_{1-42}$ oligomers do not stimulate the COX-2 expression in human microglial cell primary cultures (Hoozemans *et al.*, 2002) and microglial cells around amyloid plaques display COX-1 immunostaining (Yermakova et al., 1999). Eriksen et al., (2003) showed that non selective COX-1/2 inhibitors reduce $A\beta_{1-42}$ peptide in mice expressing human APP gene by targeting γ-secretase activity. Overexpression of COX-3, a splice variant of COX-1 and putative target of acetaminophen (Chandrasekharan *et al.*, 2002) has also been reported in AD hippocampus and $A\beta_{1-42}$ treated neuronal primary culture neuronal cells (Cui et al., 2004). Boutaud et al., (2002) demonstrated that PGH2 prostaglandin favors the *in vitro* oligomerization of $A\beta_{1-42}$ peptide and form levuglandin-adducts with the peptide, which indicates that prostaglandins might influence AD process in several ways.

PGH2 is mainly converted into PGE2 and PGD2 in brain by various prostaglandin E and D synthases. The main β-secretase BACE-1 also cleaves the membrane-bound prostaglandin E2 synthase-2 (mPGES-2) in its N-terminal end, which results in enhancement of PGE2 production (Kihara *et al.*, 2010). PGE2 produced in neuronal cells in such a way can amplify neuroinflammation since PGE2 was shown to induce IL-6 production in astrocytes (Fiebich *et al.*, 1998). By contrast, the microsomal PGE2 synthase-1 directly regulates microglial proliferation around amyloid plaques since its gene deletion in the TG2576 murine AD model decreases this proliferation and attenuates the cognitive impairments of the transgenic mice (Akitake *et al.*, 2013). PGD2 is overproduced in neuronal cells of AD patients (Iwamoto *et al.*, 1989) as well as in glial cells surrounding the amyloid plaques which overexpress the hematopoietic type of PGD2 synthase and the PGD2 receptor DP1 (Mohri *et al.*, 2007). The various production sites and targets of PGE2 and PGD2 illustrate the fact that several types of cooperation between neuronal and glial cells contribute to the extension of AD. The binding of the other lipocalin-type PGD2 synthase to Aβ peptide in CSF must be reminded. This enzyme was considered as a chaperone for Aβ peptide (Kanekiyo *et al.*, 2007; Lovell *et al.*, 2008; Hansson *et al.*, 2009), although it has probably no relationship with ARA conversion into PGD2.

Leukotrienes are the second group of eicosanoids. Their production is initiated by the conversion of free ARA by 5-lipoxygenase (5-LOX) into 5-hydroxyperoxyeicosatetraenoic acid [5-HPETE], similarly to the synthesis of PGH2 by COX-1/2. 5-HPETE will be then stabilized into 5-hydroxyeicosatetraenoic acid (5-HETE) or converted by 5-LOX into leukotriene A4 (LTA4), which is used as a substrate either by LTA4 hydrolase to give leukotriene B4 (LTB4), or by glutathione *S*-transferase to produce leukotriene C4 (LTC4). LTC4 can be further converted into leukotrienes D4 and E4 (LTD4, LTE4). In addition, the presence of an accessory 5-LOX activating protein (FLAP) is required for the optimal conversion of ARA by 5-LOX without direct interaction between the two proteins (Vickers, 1995). Ikonomovic et al., (2008) studied a small series of 11 AD and 5 control patients and described an overexpression of 5-LOX in neuronal and glial cells of the temporal cortex. But most evidences of the involvement of 5-LOX in AD were provided by studies using either 5-LOX pharmaceutical inhibitors or 5-LOX knockout mice. As soon as 1994, Goodman et al., (1994) showed that two 5-LOX inhibitors, nordihydroguaiaretic acid and AA861, protect rat hippocampal neuronal primary cultures against Aβ neurotoxicity and decrease free Ca^{2+} influx and ROS production. The administration of zileuton, another 5-LOX inhibitor, to transgenic AD model showed that 1) 5-LOX increases Aβ production by activating the CREB transcription factor and by enhancing the transcription of the constituents of the γ-secretase complex through activation (Chu & Pratico, 2011a, b), 2) 5-LOX also stimulates cdk-5 activity and expression levels, which induces Tau hyperphosphorylation (Chu & Pratico, 2013; Chu *et al.*, 2013b, c). In these studies, the administration of zileuton attenuated both amyloid and Tau pathologies observed in these murine AD models and improved their cognitive deficits. Pratico's group also demonstrated that the specific FLAP inhibitor MK-591 reduces Aβ overproduction and Tau hyperphosphorylation in the TG2576 mouse (Chu & Pratico, 2012; Chu *et al.*, 2013a), although this AD model is mainly based on the expression of the human Swedish mutant APP gene. This group confirmed the involvement of 5-LOX and FLAP in Aβ production and cdk5-mediated Tau phosphorylation by overexpressing 5-LOX gene (Chu *et al.*, 2012a, 2012b, 2013c) or by deleting the 5-LOX and FLAP genes in transgenic AD model mice (Firuzi *et al.*, 2008; Puccio *et al.*, 2011; Giannopoulos *et al.*, 2013a,b; Joshi et al., 2013a). However, by treating 3X-transgenic mice by LPS, they evidenced that 5-LOX gene deletion attenuates the worsening effect of LPS-induce neuroinflammation on Aβ production but not that on Tau hyperphosphorylation and aggregation (Joshi *et al.*, 2013b). Although the previous data were reported by the same groups, some other groups used other pharmaceutical inhibitors and concluded to the involvement of 5-LOX and FLAP in AD process. Valera et al., (2013) used a pyrazole called CNB-001 which lowers Aβ production in AD transgenic mice and suggested that 5-LOX mediates endoplasmic reticulum dysfunction. Hawkes et al., (2014) treated mice expressing a double mutant APP gene with the MK886 FLAP inhibitor and observed reduced expression levels of nicastrin, a member of the γ-secretase model. 5-LOX overexpression and LTB4 overproduction were also found in peripheral blood mononuclear cells of AD patients and were associated with reduced methylation of 5-LOX promoter DNA (Di Francesco, 2013). Furthermore, protective effects of 12-lipoxygenase inhibition have also been reported, but only on cortical neuronal cell primary cultures (Lebeau *et al.*, 2001, 2004). Pratico's group evidenced that LTB4 increases Aβ production and γ-secretase expression level in neuronal cell primary cultures (Joshi *et al.*, 2014). But Wang et al., (2013b) demonstrated that LTD4 also stimulates Aβ production in neuronal cell cultures through enhancement of β- and γ-secretase activities and these effects depends on the activation of the NF-κB signaling pathway. Therefore, since 5-LOX catalyzes the upstream reaction in the leukotriene cascade synthesis, other works are required to identify the nature of the ARA derivative(s) (leukotriene(s) or HETE(s)) which is(are) the *in vivo* causative agent(s) of these various effects.

Endocannabinoids are not direct derivatives of ARA, on the contrary to prostaglandins or leukotrienes, but ARA is one component of these complex molecules. Endocannabinoids include two signaling lipids, N-arachidonoylethanolamide (AEA) and 2-arachidonoylglycerol (2-AG). AEA is synthetized in two steps, 1) the transfer of ARA from an 1-arachidonic acid phosphatidylcholine to the polar end of a phosphatidylethanolamine with the formation of an amino-ester bound, 2) the cleavage of this lipid by a specific phospholipase D which releases AEA. 2-AG is also produced in two steps 1) the hydrolysis of a 2-arachidonyl-PI by a PI-specific phospholipase C, 2) the hydrolysis of the resulting diacylglycerol by a diacylglycerol lipase (Wang & Ueda, 2009). Only a few studies were devoted to endocannabinoid concentration in AD. No variation of circulating endocannabinoid concentration was found (Koppel *et al.*, 2009), but significantly lower levels of the endocannabinoid anandamide and its precursor 1-stearoyl, 2-docosahexaenoyl-sn-glycero-phosphoethanolamine-N-arachidonoyl (NArPE) were recently measured in frontal and temporal cortex of 38 AD patients compared to 17 control subjects without any difference for 2-AG (Jung *et al.*, 2012). By contrast, protective effects of endocannabinoids against Aβ neurotoxicity were reported in numerous studies. The first studies used neuronal cell primary cultures or on neuroblastoma cell lines to show *in vitro* neuroprotection by cannabidiol, and the two endocannabinoids (Milton, 2002; Esposito *et al.*, 2006a, b; Noonan *et al.*, 2010; Chen *et al.*, 2011, Janefjord *et al.,* 2013). Several of these studies involved the CB1 receptor, one of both G-protein coupled receptors of endocannabinoids, in this neuroprotective effects (Milton, 2002; Noonan *et al.*, 2010; Chen *et al.*, 2011). But *in vivo* studies yielded equivocal results about the role of CB1 receptor in the protection against Aβ peptide. Pamplona et al., (2012) showed that lipoxin A4, a product of ARA conversion by 15- and 5-lipoxygenase, enhanced the binding of anandamide to CB1 and decreased the Aβ-induced cognitive alterations in mice. ACEA, a specific CB1 agonist, decreased inflammation around amyloid plaques in a double mutant APP/PS1 mouse line (Aso *et al.*, 2012) and preserved the electrophysiological functions of rat hippocampus pyramidal neurons after cortical Aβ injections (Aghani *et al.*, 2012). However, Stumm et al., (2013) showed by breeding mutant APP–bearing with CB1 deficient mice that reduction of CB1 expression reduces amyloid load and inflammation, but worsens cognitive deficit. In addition, measurement of CB1 expression led to contradictory results in AD patient brains or in AD model mice (Westlake *et al.*, 1994; Lee *et al.*, 2010; Kalifa *et al.*, 2011; Mulder *et al.*, 2011; Kärkkäine *et al.*, 2012; Ahmad *et al.*, 2013). On the contrary to CB1, CB2 receptor was found to be overexpressed in AD microglial cell in several studies (Benito *et al.*, 2003; Nunez *et al.*, 2008) and its stimulation by specific agonists have positive effects on amyloid load, neuroinflammation, tauopathy and cognitive abilities in AD model mice (Martin-Moreno, 2012; Solas *et al.*, 2013; Wu *et al.*, 2013; Aso *et al.*, 2013). Another endocannabinoid-associated neuroprotective mechanism was indicated by Bachmeier et al., (2013) who found that endocannabinoids increase LRP-1 expression level in BBB and subsequently the clearance of Aβ peptide. Only a few studies compared the respective effects of 2-AG and AEA, the first one being quantitatively the major endocannabinoid in brain. AEA, but not 2-AG, increases the expression of nicastrin and favors the Notch signaling pathway, thereby promoting the $A\beta_{1-40}$ over the $A\beta_{1-42}$ production in neuronal cell primary cultures (Tanver *et al.*, 2012). On the contrary, only 2-AG protects the human neuroblastoma SH-SY5Y cell line against the Aβ-induced apoptosis (Janefjord, 2013). Whether the discrepancy between these two studies might be due to the different cell models remains to be determined.

The F2 isoprostanes are the last group of ARA derivatives which can be measured in AD patients. These compounds result from the oxidation of ARA, which also affects other PUFAs, mainly DHA which leads to neuroprostanes [or F4-isoprostane]. Higher levels of both types of compounds were measured in AD brain as well as in other neurodegenerative diseases (Montine *et al.*, 2004). Moreover, ARA and DHA undergo oxidation in membrane phospholipids (Morrow *et al,*. 1992) and are released by type-II secretory phospholipases A_2 and PAF acetyl hydrolases (Tselepis & Chapman, 2006). They correspond to the non-specific oxidative stress associated to AD as to other neurodegenerative disease and do not seem to be causative agents of AD.

We have seen that several studies reported overexpression of enzymes converting ARA into prostaglandins or leukotrienes in patients or in AD model mice, although there are some conflicting results in patients, probably due to the restricted size of the series and to the difficulty to handle *post-mortem* samples. This overexpression indicates the involvement of these enzymes and their products in AD (Figure 3). Studies on cellular models or on transgenic mice indicate that these pathways participate not only to the extension of the neuroinflammatory processes, but also to the key mechanisms of AD, i.e., Tau hyperphosphorylation and Aβ overproduction or reduction of its clearance. Taken together, these studies provided the preclinical basis for testing 5-LOX or COX inhibitors in treatment of AD patients. But conventional NSAIDs did not display any efficiency against AD in clinical trials, at least after the establishment of the disease. Therefore, clinical studies are required to evaluate the early interest of COX-2 or 5-LOX specific inhibitors. On the contrary to eicosanoids, endocannabinoids display positive effects on neuroinflammation and key AD mechanisms on cell and AD model mice. Additional studies are necessary to determine the respective contribution of the pathways depending on CB1 or CB2 receptors and a better elucidation of these pathways. AD therapy could benefit of the finding of new specific agonists devoid of psychomimetic effect. Eicosanoid synthesis is initiated by the release of free ARA which can be considered as a key step for the production of the various prostaglandins and leukotrienes. Furthermore, endocannabinoids degradation generates free ARA. Therefore, limitation of release of free ARA could have beneficial effect not only to treat but also to prevent AD. The relationship between AD and the enzymes releasing ARA from membrane phospholipids and endocannabinoids will be examined below.

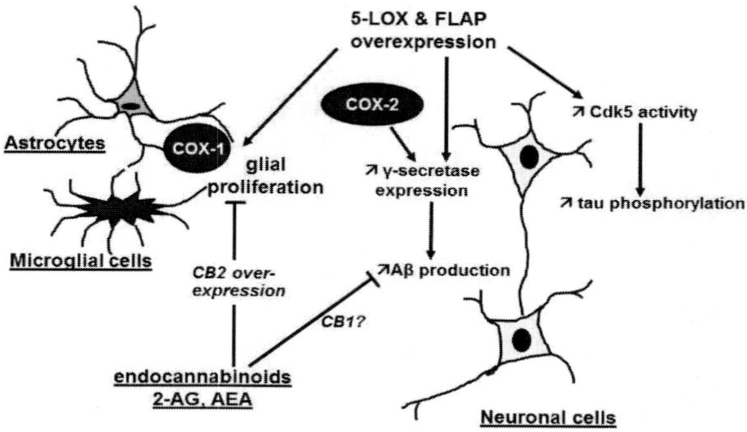

Figure 3. Roles of 5-LOX/FLAP and COX1/2 on AD mechanisms. Endocannabinoids counteract the eicosanoid effects. Overexpression and role of CB2 receptors in glial cells have been reported in several studies whereas conflicting results have been published about CB1 expression levels and role in AD.

SOURCE OF FREE ARACHIDONIC ACID IN BRAIN AND ALZHEIMER'S DISEASE

The main cellular source of free ARA is the release of this PUFA from membrane phospholipids by phospholipases A_2. These proteins constitute a superfamily which includes intracellular as well secretory and Ca^{2+}-dependent or -independent enzymes (Burke & Dennis, 2009). Before 2000, total phospholipase A_2 activities were measured and found reduced in brain cortex of AD patients (Gattaz *et al.*, 1995), but no indication about the identity of the enzymes was provided. In a more recent period, immunoreactivity for the secretory phospholipase A_2 type IIA ($sPLA_2$-IIA) was observed in astrocytes of AD patient hippocampus and inferior temporal gyrus with a stronger intensity in the astrocytes around amyloid plaques (Moses *et al.*, 2006). These data are consistent with the role of $sPLA_2$-IIa in inflammation (Burke & Dennis, 2009) and its expression in astrocyte cultures (Oka & Arita 1991). Indeed, these authors confirmed that interleukin-1 and $A\beta_{1-42}$ peptide induce the expression of $sPLA_2$-IIA mRNA in human astrocyte cultures (Moses *et al.*, 2006). Treatments of neuronal cell cultures by exogenous ARA or another secretory phospholipase A_2, PLA_2 type III, increased the localization of APP at the membrane surface, its degradation by α-secretase and reduced the Aβ production (Yang *et al.*, 2010). These data were reminiscent to those of previous studies about the processing of APP in CHO cells expressing this protein and the m1 muscarinic receptor (Emmerling *et al.*, 1993, 1996). But these secretory PLA_2 release all the fatty acids from the sn-2 position of membrane phospholipids without any specificity for ARA. A calcium-dependent cytosolic phospholipase A_2, named $cPLA_2$-α or $cPLA_2$ type IVA according Dennis & Burke's classification, is the only phospholipase A_2 which specifically used as substrate phospholipids bearing ARA at their sn-2 position. Translocation of this cytosolic enzyme to membranes requires phosphorylation on its Ser^{505} by the MAP-kinase cascade although phosphorylation of other sites by other kinases has been observed (Hirabayashi *et al.*, 2004). Stephenson et al., (1996) suggested the involvement of $cPLA_2$-α in AD by describing a stronger immunoreactivity of this protein with a monoclonal antibody in 5 AD patient cortex compared to 6 control subjects. This immunoreactivity colocalized mainly with GFAP with a higher intensity around amyloid plaques, indicating an expression of $cPLA_2$-α in astrocytes and its association with neuroinflammation. This result was consistent with the previously described expression of $cPLA_2$-α in human astrocytes (Stephenson *et al.*, 1993). Microglial cells also contribute to $cPLA_2$-α overexpression in human and mice in several brain pathologies including AD (Stephenson *et al.*, 1999). The increased expression of $cPLA_2$-α in AD hippocampus was confirmed by Lukiw (2004) in mRNA microarrays experiments and was related to the overexpression of other neuroinflammatory markers including COX-2. The association between $cPLA_2$-α and AD process in glial cells was thereafter supported by several studies on astrocytes and microglial cell cultures (Zhu *et al.*, 2006; Hicks *et al.*, 2008; Szaingurten-Solodkin *et al.*, 2009). These works established that $A\beta_{1-42}$ peptide-induced activation of NADPH oxidase in these cells leads to ROS production and $cPLA_2$-α phosphorylation and translocation to membranes, peculiarly to mitochondria membranes which play a major role in oxidative stress (Zhu et al., 2006). In addition, $A\beta_{1-42}$ peptide also increases expression of $cPLA_2$-α and COX-2 through the activation of the redox-sensitive NF-κB transcription factor, at least in microglial cells (Szaingurten-Solodkin, 2009). Reciprocally, inhibition of the cPLA$_2$-α expression decreases NADPH oxidase activation, NO production and modifications of astrocyte membrane (Szaingurten-Solodkin, 2009; Hicks *et al.*, 2009). The strongly linked activations of cPLA$_2$-α and NADPH oxidase by Aβ$_{1-42}$ peptide suggest a common upstream step in the initiation of the Aβ$_{1-42}$ signaling pathway. Askarova et al., (2011) reported that the receptor for advanced glycation endproducts [RAGE] binds Aβ$_{1-42}$ peptide on the plasma membranes of astrocytes and endothelial cells and triggers the downstream events. RAGE has been previously described as a major actor in the mediation of Aβ$_{1-42}$ peptide effects and its homeostasis in brain (for review, see Deane *et al.*, 2004). Activation of cPLA$_2$-α and MAP-kinase cascade are triggered in retina pericytes like in endothelial cells, which underlines the vascular effects of Aβ oligomers (Anfuso *et al.*, 2004, Nicotra *et al.*, 2005).

Although it has not been initially noted, cPLA$_2$-α is also expressed in neuronal cells in which it mediates the neurotoxic effects of Aβ$_{1-42}$ peptide. In 2005, Kriem et al., (2005) demonstrated for the first time that the reduction of the cPLA$_2$-α activity or expression can protect the neuronal cells in primary cultures from the Aβ peptide-induced apoptosis. The cPLA$_2$-α activation pathways induced by Aβ peptide were investigated and the critical role of the MAP-kinase cascade was evidenced by using Erk1/2 and p38 inhibitors. These results were obtained by using either Aβ$_{1-40}$ or Aβ$_{1-42}$ oligomers although the last type is considered as much more toxic. Bate & William (2010) described that Aβ$_{1-42}$ peptide decreases the expression level of synaptophysin, the recycling of synaptic vesicles in mouse cortical neuronal cell cultures and phosphorylation of cPLA$_2$-α. This illustrates the synaptotoxicity of this peptide in a cell model and the involvement of cPLA$_2$-α which was considered by the authors as a presynaptic marker. These authors also observed that when previously mixed with Aβ$_{1-42}$, Aβ$_{1-40}$ peptide decreases Aβ$_{1-42}$ effects including cPLA$_2$-α activation. The discrepancy between the results of the two studies regarding Aβ$_{1-40}$ effect could be related to the fact that the first study described neurotoxic effects of Aβ$_{1-40}$ at 1 μM concentrations while the second one found neuroprotective at ten-fold lower concentrations. As in glial cells, NADPH synthase regulates cPLA$_2$-α activity in response to Aβ peptide in neuronal cell culture and NADPH oxidase is controlled by NMDA receptors in these neuronal cells since antagonists of both NMDA receptors and NADPH oxidase decrease the Aβ-induced release of ARA and cPLA$_2$-α phosphorylation (Shelat *et al.*, 2008). This work confirmed that MAP kinases are activated by Aβ peptide in neuronal cells as previously reported (Kriem *et al.*, 2005). Expression of wild-type or mutant human APP in neuroblastoma PC12 cells, which produces intracellular Aβ peptide, leads to activation of neutral NO synthase, $cPLA_2$-α phosphorylation and ARA release. This ARA release was reduced by NO synthase and γ-secretase inhibitions, indicating that the neuronal $cPLA_2$-α dependent release of ARA occurs in the frame of an inflammatory process that includes the usual partners of this reaction (NO synthase, NADPH oxidase). $cPLA_2$-α is also upregulated in neuronal cells from p25 overexpressing mice (Sundaram *et al.*, 2012). This finding is noticeable since p25 activates cdk5, which was involved in the dysregulated hyperphosphorylation of Tau. This result can be linked to those of another group which used the MAFP as *in vitro* and *in vivo* $cPLA_2$-α inhibitor (Schaeffer *et al.*, 2011). However, the *in vivo* specificity and efficiency of MAFP are questionable, since MAFP also inhibits calcium–independent PLA_2 and the endocannabinoid hydrolyzing enzymes [see below]. More specific and potent inhibitors of $cPLA_2$-α should be used to confirm these data. Such $cPLA_2$-α inhibitors have been rarely used in animal assay and might be unstable at least for some of them (Fabian *et al.*, 2013).

The way by which $cPLA_2$-α-released ARA acts to contribute to AD process was investigated in several studies. COX-2 inhibition limits the pro-apoptotic effects of Aβ oligomers by 50%, but the 5-LOX putative contribution was not explored by the authors (Kriem *et al.*, 2005). High levels of free ARA, several epoxyeicosatrienoic; diepoxyeicosatrienoic acids and 12-HETE, but no other prostaglandins and leukotrienes, were measured in brain hippocampus and cortex of mice expressing mutant APP gene (Sanchez-Meija *et al.*, 2008). High levels of phosphorylated $cPLA_2$-α were found in the hippocampus of these AD model mice as well as in the hippocampus of AD patients. By contrast, lipidomics experiments found that lysophosphatidylcholine, the other product of the phosphatidylcholine hydrolysis by $cPLA_2$-α, initiate neuroinflammation and astrogliosis induced by p25 overexpression (Sundaram *et al.*, 2012). Activation of neutral and acid sphingomyelinases is another rapid downstream event of $A\beta_{1\text{-}42}$-induced $cPLA_2$-α activation in neuronal cell primary cultures, which could be inhibited by sphingosine-1 phosphate (Malaplate-Armand *et al.*, 2006). A single intracerebroventricular injection of 1 nanomole of $A\beta_{1\text{-}42}$ oligomers activates neutral and acid sphingomyelinases and drastically reduces Akt/PKB phosphorylation and expression of synaptic markers (Desbène et al., 2012). As a result, short- and long term memory abilities of the treated mice were deeply altered within the 2-week period following $A\beta_{1\text{-}42}$ injection (Desbène *et al.*,

2012). Neuronal activation of cPLA$_2$-α would increase the localization of the non-NMDA glutamate AMPA receptor at the membrane surface and this would be the way that this enzyme contributes to Aβ neurototoxicity (Sanchez-Meija *et al.*, 2008). Breeding cPLA$_2$-α$^{-/-}$ and mutant APP gene expressing mice attenuate the memory deficits, reduce the eicosanoid concentration in cortex and hippocampus and increase the survival of the resulting mice compared to the AD parent mice (Sanchez-Meija *et al.*, 2008). Interestingly, APP expression is reduced in hippocampus and neuronal cells of cPLA$_2$-α$^{-/-}$ mice. Taken together with the resistance of these mice to Aβ$_{1-42}$ neurototoxicity and the fact that APP expression induce sensitivity to the Aβ$_{1-42}$ oligomers in HEK293 cells (unpublished results, this result suggests that APP could be a receptor for these oligomers. Shaked et al., (2006) previously reported that Aβ$_{1-42}$ oligomers interact with APP extracellular domain.

The degradation of endocannabinoids, after their reuptake by neuronal cells, is the second source of ARA. On one hand, the fatty acid amide hydrolase (FAAH) hydrolyses all the N-acylethanolamines, including AEA which is a minor product in this family (Wang *et al.*, 2009; Benito *et al.*, 2012). On the other hand, the major endocannabinoid 2-AG is mainly degraded in brain by monoacylglycerol lipase (Schlosburg *et al.*, 2010, Pan *et al.*, 2011). Higher expression and activity of FAAH was found in AD patients in association to a decrease in DNA methylation of FAAH promoter (D'Adario *et al.*, 2012). Therefore, one could expect that reduction of FAAH expression or activity attenuates inflammation in patients or AD cell and animal models. However, Benito et al., (2012) observed that astrocytes from FAAH$^{-/-}$ mice are highly responsive to 5 μM Aβ$_{1-42}$ oligomers in terms of induced expression of IL-1β, COX-2, NO synthase, NF-κB and other inflammatory markers. AEA as well as other N-acylethanolamine partially reversed the Aβ$_{1-42}$–induced proinflammatory effects. But the authors failed to block the induction of inflammatory markers by Aβ$_{1-42}$ oligomers by using CB1 or CB2 receptors antagonist. By contrast, they found increased mRNA levels of the transient receptor potential vanilloid-1 [TRPV1] which binds N-acylethanolamines. In addition, treatments of wild-type astrocytes by pharmaceutical FAAH inhibitors did not reproduce an inflammatory phenotype. These results suggest that the inflammatory reaction observed in FAAH$^{-/-}$ astrocytes in presence of Aβ$_{1-42}$ oligomers results from an adaptive phenotype and the actions of several N-acylethanolamines and not only from the lack of AEA degradation. No other study on the consequences of modification of FAAH activity in AD cell or animal models has been published until now to our knowledge. Regarding the second endocannabinoid-degrading enzyme, the MAGL$^{-/-}$ mice exhibit large modifications of synaptic functions, some of them being related to CB1 desensitization by high levels of 2-AG (Pan *et al.*, 2011). The MAGL$^{-/-}$ mice display better abilities in learning tasks compared to the wild type animals (Pan *et al.*, 2011). Chen et al., (2012) used 5X-FAD mice bearing 3 and 2 mutations of the human APP and PS1 genes, respectively, and administrated intraperitoneally the MAGL inhibitor JZL184 for 8 weeks in 4-month old animals. This treatment reduced by more than two third the total surface of amyloid plaques in cortex and hippocampus, drastically decreased the $A\beta_{1-42}$ production in cortex and at a lower extent in hippocampus. It also normalized the expression of BACE-1 which was increased in transgenic mice compared to the control animals. In addition, JZL184 administration preserved the synaptic structure and expression of NMDA and AMPA receptors in transgenic mouse hippocampi. However CB1 or CB2 receptors do not mediate the effect of MAGL inhibition since animals resulting from the breeding of CB1 or CB2 KO were still sensitive to JLZ184 treatment. Another group (Piro *et al.*, 2012) showed that MAGL gene suppression in mice leads to enhancement of 2-AG and drastic reduction of free ARA brain concentrations. The eicosanoids PGE2, PGD2 and TXB2 were also drastically reduced in MAGL$^{-/-}$ mouse brains. Furthermore, lipidomic analysis of PAD PS1/APP mouse brains revealed high concentrations of free ARA, PGE2, PGD2 and TXB2 as well as a reduction of 2-AG associated with a compensation by AEA. Breeding these animals with MAGL$^{-/-}$ mice suppressed the enhancement of ARA and eicosanoids concentrations induced by APP/PS1 mutations and normalized the 2-AG concentrations compared to the MAGL$^{-/-}$ mouse brains. MAGL inactivation also decreased Aβ production, amyloid plaque formation and neuroinflammation associated with APP/PS1 mutations. Finally, treatments by JZL184 or a CB1/CB2 receptors antagonist for 16 days suppressed production of pro-inflammatory cytokines and eicosanoids in brain of 6-month old APP/PS1 mice. Taken together, these results showed that increased degradation of 2-AG orientates free ARA to the synthesis of pro-inflammatory eicosanoids, which amplifies neuroinflammation in AD mice. Lastly, evidence about a role for MAGL in AD patients were provided by a third group (Piro *et al.*, 2012) who showed that MAGL expression correlates with Braak stages in AD patients and AD induces MAGL relocalization from cortical membranes to cytosol. These authors also reported that the expression of another 2-AG degrading enzyme, the ABHD6 serine hydrolase, was increased in microglia around amyloid plaques and decreased in hippocampal pyramidal neurons of AD patients.

All these results established that cPLA$_2$-α and MAGL are the main enzymes providing free ARA in glial and neuronal cells and play a putatively major role in AD. These enzymes are overexpressed and/or activated in hippocampal and cortical regions of AD patient brains. Experiments with pharmaceutical inhibitors in wild-type and transgenic mice also supported the involvement of these enzymes. However, additional studies are required to elucidate the ways that cPLA$_2$-α and MAGL cooperate with the downstream eicosanoids enzymes in the synthesis of the proinflammatory prostaglandins and leukotrienes. Furthermore, whether and how ARA could play a role without being converted into eicosanoid, remains to be investigated. The contribution of lysophosphatidylcholine or other lysophospholipids resulting from the hydrolysis of membrane phospholipids by cPLA2-α is still an open question.

ARACHIDONIC ACID ACTIONS AND HOMEOSTASIS IN BRAIN AND ALZHEIMER'S DISEASE

ARA binds and activates several protein kinases such as PKN (Mukai et al., 2003) and PKCξ (Kochs *et al.*, 1993). PKN family is constituted by 3 PKC-homologous serine/protein kinases, the first one, PKNα, being highly expressed in brain (Mukai, 2003). In addition, PKN family members are activated at least *in vitro* by ARA or linoleic acid and by Rho family GTPases that were identified as important regulators of the actin cytoskeleton (Mukai, 2003). In agreement with their relation by Rho GTPases, PKNα is involved in the regulation of neuronal microtubules and phosphorylates Tau (Kawamata *et al.*, 1998). Accumulation of PKNα was found in neurofibrillary tangles (Taniguchi *et al.*, 2001). However, the role of free ARA in putative dysregulation of PKN has not been determined until now in AD or other neurodegenerative pathologies. Excessive striated muscle growth in the young cPLA2-$\alpha^{-/-}$ mice has been associated to the inactivation of PKCξ and the suppression of its repressive effect on the IGF-1 signaling pathways. These results and their conclusion are of interest for AD process since IGF-1 has a positive action in AD and has been tested in mice (Zemva & Schubert, 2013). Activation of PKCξ by excessive intraneuronal concentrations of free ARA would potentially inhibit IGF1 pathway. Increased levels of PKCξ expression have been measured in neuronal membrane fraction of AD TG2576 mice (Rossner *et al.*, 2001) and in a T-lymphocyte subpopulation of severe AD patients (Miscia *et al.*, 2009). Furthermore, PKCξ contributes to Tau phosphorylation by targeting leucine-rich repeat kinase 2 (LRRK2) whose mutation is the most frequent cause of genetic forms of Parkinson disease (Zach *et al.*, 2010). Besides the protein kinase activation, the other main neuronal process modulated by ARA is the formation of SNARE complex and the resulting exocytosis of neuromediators. ARA induces SNARE formation by interacting with syntaxin-1A and munc18 (Rickman & Davletov, 2005; Latham *et al.*, 2007; Connel *et al.*, 2007). Activation of SNARE and exocytosis can be blocked through the sequestration of ARA by α-synuclein whose mutations are implicated in the genetic forms of Parkinson disease (Darios *et al.*, 2010). Furthermore, syntaxin-1 and munc18 also interact with XL11 and APP to segregate APP in peculiar membrane microdomains which differ from those containing BACE-1 (Sakurai *et al.*, 2008). Phosphorylation of munc18 by cdk5 induces a shift of APP to BACE-1-containing domain (Sakurai *et al.*, 2008). Munc18 can also be phosphorylated by the dual-specificity tyrosine[Y]-phosphorylation-regulated kinase 1A [Dyrk1A] which might be responsible for the early onset of AD in Down's syndrome patients (Park *et al.*, 2012). Whether intraneuronal free ARA could modulate the interaction between syntaxin-1A, Munc18 and APP has not been investigated in these studies. Munc18, XL11 and cdk5 were found upregulated in the temporal and occipital cortex of AD patients but not in TG2576 mice (Jacobs *et al.*, 2006). This upregulation only in humans has been assumed as a compensatory mechanism to maintain synaptic functions and avoid excessive production of Aβ peptide in surviving neurons.

The actions of free ARA (Figure 4) and its cellular fate depend on its uptake by several fatty acid-binding proteins (FABP) which also transport other fatty acids to different organelles and/or enzymes. Three FABPs have been characterized in brain until now (Liu *et al.*, 2010). FAPB7 (or brain FABP, B-FABP) contributes to glioma infiltration in presence of high ARA/DHA ratios (Mita *et al.*, 2010). FABP5 (or epidermal FABP, E-FABP) favors neurite growth in neuroendocrine PC12 cells in presence of neuronal growth factor and ARA, DHA or eicosapentaenoic acid (Liu *et al.*, 2008). Deletion of FABP3 (or heart FAPB, H-FABP) gene reduces ARA distribution by 25% in phosphatidylinositol, which is noteworthy since phosphatidylinositol contains high amounts of ARA in brain as well as in other tissues (Tanaka *et al.*, 2003) and ARA turnover is especially high in this phospholipid group in the neuronal growth cone (Negre-Aminou & Pfenninger, 1993). Activation of fatty acid by various acyl-CoA synthetases (ACSLs) is critical for their incorporation in membrane phospholipids and triglycerides. Among the various ACSLs, ACSL4 specifically uses ARA as substrate and was originally characterized in steroidogenic tissues (Kang *et al.*, 1997). Thereafter, its mutation was associated to X-linked mental retardation (Piccini *et al.*, 1997; Meloni *et al.*, 2009). Two isoforms of ACSL4 are expressed in neuronal cells including a neuron-specific one with an additional 41 amino-acid N-terminal domain (Meloni *et al.*, 2009: Cho *et al.*, 2012). ACSL-4 expression increases during neuronal cell differentiation and neurite growth (Meloni *et al.*, 2009: Cho *et al.*, 2012). The reduction of its expression by silencer RNAs alters the dendritic spine architecture in rat hippocampal neurons (Meloni *et al.*, 2009). Mutations of its analog in drosophila alter the axonal transport of vesicles and synaptic development (Liu *et al.*, 2011). These results suggest that ACSL4 could play a role in AD in which synaptic dysfunctions are the first alterations. Furthermore, overexpression of ACSL4 reduced basal and IL-1β-induced PGE2 secretion in vascular smooth muscle cells (Golej *et al.*, 2011). Conversely, acute pharmaceutical inhibition of ACSL4 by triacsin C enhanced PGE2 release in this study without upregulating COX-2 expression. But the authors also observed that long-term reduction of ACSL4 expression by stably expressed siRNA also reduces PGE2 production and underline that this sustained downregulation affects cell growth in vascular smooth muscle cells like in tumor cells. ACSL4 expression has not been reported in glial cells, but the mutation of it drosophila paralog drastically reduces the number of glial and neuronal cells in the drosophila brain development (Zhang *et al.*, 2009). The expression of ACSL4 in human glial cells could limit prostaglandin production and have some influence on the extension of neuroinflammation in AD. The other main result reported in the role of ACSL4 in vascular smooth muscle cells is that overexpression of ACSL4 increased ARA incorporation by 3-fold into phosphatidylethanolamine and phosphatidylinositol, and by 1.5-fold in triacylglycerol of vascular smooth muscle cells. By expressing several ACSL4 isoforms in CHO cells, Küch et al., (2013) showed that exogenous ARA is preferentially channeled towards phosphatidylinositol by ACSL4. Moreover, ARA incorporation into phosphatidylinositol is selectively catalyzed by lysophosphatidylinositol acyltransferase 1 (LPIAT1, also known as MBOAT7) which is a putative ACSL4 partner. Reduction of its expression drastically affects neurite growth and LPIAT1$^{-/-}$ mice display cortex and hippocampus atrophy which leads to their death within a month (Lee *et al.*, 2012).

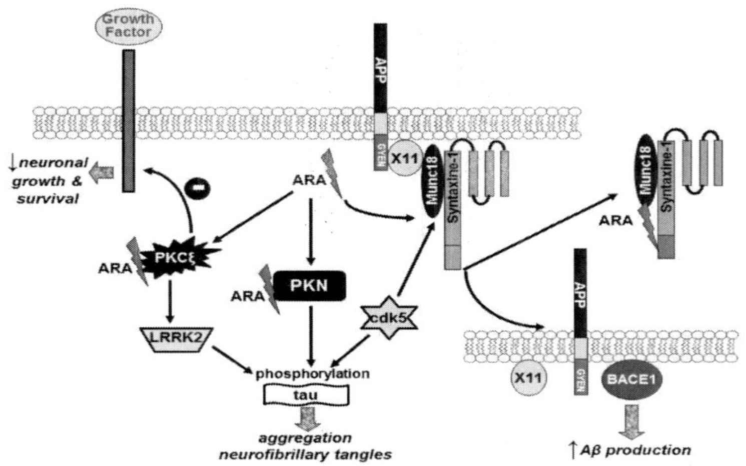

Figure 4. Putative targets and modes of action of free arachidonic acid in AD.

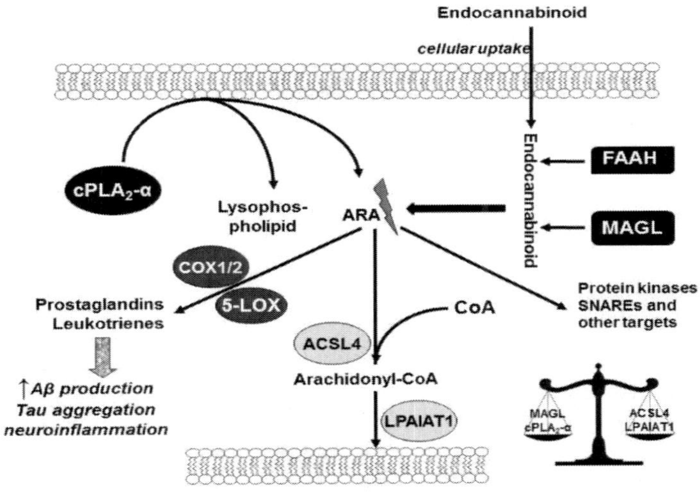

Figure 5. Balance between the enzymes which release or incorporate arachidonic acid. Free arachidonic acid levels in neuronal cells are increased through the degradation of endocannabinoids by MAGL/FAAH and the release of arachidonic acid from membrane phospholipids by $cPLA_2$-α. On the contrary, ACSL4 and LPAIAT1 or other lysophospholipids transferases decrease these levels by favoring incorporation of arachidonic acid into membrane phospholipids. Therapeutic or preventive strategies against AD could target this balance by stimulating the activities of ACSL4 and lysophospholipid acyltransferases.

Taken together, these results indicate that ACSL4 and LPAIAT1 combined action can limit free ARA levels resulting from the cPLA$_2$-α or MAGL activation and limit its conversion into neurotoxic eicosanoids (Figure 5). The other ways to limit release of free ARA into neuronal cells would be to avoid excessive concentrations of ARA in brain membrane phospholipids. Polyunsaturated fatty acids, mainly ARA and DHA, are either synthetized by astrocytes in brain, either provided by food and liver synthesis. Diets in Western countries incorporated more and more ω-6 fatty acids in the last decades. Epidemiological studies evaluate that ω-6/ω-3 precursor ratio increased from 6 to 10 and more in USA between the beginning and the end of the last century (Blasbalg *et al.*, 2011). The situation should be close in the other high-income countries and probably in many mid or low-income counties where obesity prevalence increases. A tremendous number of studies has been devoted to the influence of deficiencies or supplementations of dietary ω-3 fatty acids on cognitive functions in humans and animals (for review, see Luchtman & Song, 2013). But a little number of works studies the influence of excessive amount of ARA or ω-6 precursors in food. High amounts of linoleic and ARA in food to piglets reduce their cerebral accretion of DHA and compromise neurite growth (Novak *et al.*, 2008). A diet enriched with 2% ARA given between 3 to 24 weeks of age was associated with an increase in number and size of amyloid plaques in cortex and hippocampus of double mutant APP mice as compared to the standard diet-fed animals (Amtul *et al.*, 2012). Cognitive abilities in a spatial learning task were compared in young and old rats in association with the fatty acid contents of various brain regions (Yetimler *et al.*, 2012). This study confirmed low cognitive abilities in old rats compared to the young group. But old rats displayed large interindividual variations and good/poor learners could be distinguished in each group. No correlation was observed in both group between DHA brain and cognitive abilities. Young good learners had lower ARA amounts in hippocampus than young poor learners, but no difference was observed in the old group. The results of these last two groups argue in favor of the benefit of low amounts of ARA in food. But positive effects of ARA supplementation have been reported in the alleviation of age-related deficiencies of various brain functions in animals (Okaichi *et al.*, 2005; Fukaya *et al.*, 2007). Therefore, additional studies are necessary to evaluate the optimal dietary amounts of ARA and ω-6 precursors in humans at various ages to prevent and decrease the occurrence of the age-related neurodegenerative disease such as AD.

CONCLUSION

ARA is one of the two main PUFAs in brain with DHA. ARA through direct effects or upon conversion into various eicosanoids can favor AD process and progression, not only by increasing neuroinflammation but also by enhancing synaptic and neuronal dysfunctions. It is therefore a critical actor in the vicious cycle of AD which associates synaptic alterations, neuronal degeneration, glial cell proliferation, synthesis of neuroinflammatory mediators. Free intracellular ARA is released from phospholipids or from endocannabinoid by $cPLA_2$-α or MAGL, respectively. It can be converted into prostaglandins by cyclooxygenases and prostaglandin synthases or to leukotrienes or HETE by 5-LOX and other lipoxygenases. Taking into account this complex network of reactions, therapeutic strategies targeting ARA roles should combine several inhibitors or use pluripotent molecules. The finding of such molecules and new clinical studies are necessary to determine whether this strategy can provide efficient solutions in the fight against AD. Besides the inhibition of $cPLA_2$-α or MAGL, favoring the re-incorporation of free ARA into membrane phospholipids would be an alternative way to limit its intracellular levels and to slow its conversion into eicosanoids. The interplay of ACSL4 and LPAIAT1 or other lysophospholipid acyltranferases constitutes a new putative target which should be further investigated on AD cellular and animal models. But, adequate release of ARA at the right place and at the right moment is probably necessary to allow neuronal functions and synaptic plasticity. These studies will also provide new essential informations about the regulation of ARA cellular homeostasis. Nutritional studies combined to experiments on cellular models are also required to improve the knowledge about the role of ARA in aging and AD. The determination of the adequate dietary ARA and ω-6 intake at different ages of life will be helpful to design preventive strategies against pathological aging.

REFERENCES

Ahmad, R; Goffin, K; Van den Stock, J; De Winter, FL; Cleeren, E; Bormans, G; Tournoy, J; Persoons, P; Van Laere, K; Vandenbulcke, M. *In vivo* type 1 cannabinoid receptor availability in Alzheimer's disease. *Eur Neuropsychopharmacol.* 2013, pii, S0924-977X00294-0. doi, 10.1016/j.euroneuro.2013.10.002.

Haghani, M; Shabani, M; Javan, M; Motamedi, F; Janahmadi, M. CB1 cannabinoid receptor activation rescues amyloid β-induced alterations in behaviour and intrinsic electrophysiological properties of rat hippocampal CA1 pyramidal neurones. *Cell Physiol Biochem.* 2012, 29, 391-406.

Akitake, Y; Nakatani, Y; Kamei, D; Hosokawa, M; Akatsu, H; Uematsu, S; Akira, S; Kudo, I; Hara, S; Takahashi, M. Microsomal prostaglandin E synthase-1 is induced in alzheimer's disease and its deletion mitigates alzheimer's disease-like pathology in a mouse model. *J Neurosci Res.* 2013, 91, 909-19.

Amtul, Z; Uhrig, M; Wang, L; Rozmahel, RF; Beyreuther, K. Detrimental effects of arachidonic acid and its metabolites in cellular and mouse models of Alzheimer's disease, structural insight. *Neurobiol Aging.* 2012, 33, 831.e21-31.

Anfuso, CD; Assero, G; Lupo, G; Nicotra, A; Cannavò, G; Strosznajder, RP; Rapisarda, P; Pluta, R; Alberghina, M. Amyloid beta(1-42) and its beta(25-35) fragment induce activation and membrane translocation of cytosolic phospholipase A2 in bovine retina capillary pericytes. *Biochim Biophys Acta.* 2004, 1686, 125-38.

Askarova, S; Yang, X; Sheng, W; Sun, GY; Lee, JC. Role of Aβ-receptor for advanced glycation endproducts interaction in oxidative stress and cytosolic phospholipase A_2 activation in astrocytes and cerebral endothelial cells. *Neuroscience.* 2011, 199, 375-85.

Aso, E; Juvés, S; Maldonado, R; Ferrer, I. CB2 cannabinoid receptor agonist ameliorates Alzheimer-like phenotype in AβPP/PS1 mice. *J Alzheimers Dis.* 2013, 35, 847-58.

Aso, E; Palomer, E; Juvés, S; Maldonado, R; Muñoz, FJ; Ferrer, I. CB1 agonist ACEA protects neurons and reduces the cognitive impairment of AβPP/PS1 mice. *J Alzheimers Dis.* 2012, 30, 439-59.

Bachmeier, C; Beaulieu-Abdelahad, D; Mullan, M; Paris, D. Role of the cannabinoid system in the transit of beta-amyloid across the blood-brain barrier. *Mol Cell Neurosci.* 2013, 56, 255-62.

Bate, C; Williams, A. Amyloid-β[1-40] inhibits amyloid-β[1-42] induced activation of cytoplasmic phospholipase A_2 and synapse degeneration. *J Alzheimers Dis.* 2010, 21, 985-93.

Benito, C; Núñez, E; Tolón, RM; Carrier, EJ; Rábano, A; Hillard, CJ; Romero, J. Cannabinoid CB2 receptors and fatty acid amide hydrolase are selectively overexpressed in neuritic plaque-associated glia in Alzheimer's disease brains. *J Neurosci.* 2003, 23, 11136-41.

Benito, C; Tolón, RM; Castillo, AI; Ruiz-Valdepeñas, L; Martínez-Orgado, JA; Fernández-Sánchez, FJ; Vázquez, C; Cravatt, BF; Romero, J. β-Amyloid exacerbates inflammation in astrocytes lacking fatty acid amide hydrolase through a mechanism involving PPAR-α; PPAR-γ and TRPV1; but not CB_1 or CB_2 receptors. Br J Pharmacol. 2012, 166, 1474-89.

Blasbalg, TL; Hibbeln, JR; Ramsden, CE; Majchrzak, SF; Rawlings, RR. Changes in consumption of omega-3 and omega-6 fatty acids in the United States during the 20th century. *Am J Clin Nutr.* 2011, 93, 950-62.

Boutaud, O; Ou JJ; Chaurand, P; Caprioli, RM; Montine, TJ; Oates, JA. Prostaglandin H2 tPGH2) accelerates formation of amyloid beta1-42 oligomers. *J Neurochem.* 2002, 82, 1003-6.

Brandt, R; Hundelt, M; Shahani, N. Tau alteration and neuronal degeneration in tauopathies, mechanisms and models. *Biochim Biophys Acta.* 2005, 1739, 331-54.

Burke, JE; Dennis, EA. Phospholipase A2 structure/function; mechanism; and signaling. *J Lipid Res.* 2009, 50 Suppl, S237-42.

Carlsson, CM. Type 2 diabetes mellitus; dyslipidemia; and Alzheimer's disease. *J Alzheimers Dis.* 2010, 20, 711-22.

Chandrasekharan, NV; Dai, H; Roos, KL; Evanson, NK; Tomsik, J; Elton, TS; Simmons, DL. COX-3; a cyclooxygenase-1 variant inhibited by acetaminophen and other analgesic/antipyretic drugs, cloning; structure; and expression. *Proc Natl Acad Sci U S A.* 2002, 99, 13926-31.

Chang, KA; Suh, YH. Possible roles of amyloid intracellular domain of amyloid precursor protein. *BMB Rep.* 2010, 43, 656-63.

Chen, R; Zhang, J; Wu, Y; Wang, D; Feng, G; Tang, YP; Teng, Z; Chen C. Monoacylglycerol lipase is a therapeutic target for Alzheimer's disease. *Cell Rep.* 2012, 2, 1329-39.

Chen, X; Zhang, J; Chen, C. Endocannabinoid 2-arachidonoylglycerol protects neurons against β-amyloid insults. *Neuroscience.* 2011, 178, 159-68.

Cho, YY. A novel role of brain-type ACS4 isotype in neuronal differentiation. *Biochem Biophys Res Commun.* 2012, 419, 505-10.

Choi, J; Joseph, L; Pilote, L. Obesity and C-reactive protein in various populations, a systematic review and meta-analysis. *Obes Rev.* 2013, 14, 232-44.

Chu, J; Giannopoulos, PF; Ceballos-Diaz, C; Golde, TE; Praticò D. 5-Lipoxygenase gene transfer worsens memory; amyloid; and tau brain pathologies in a mouse model of Alzheimer disease. *Ann Neurol.* 2012a, 72, 442-54.

Chu, J; Giannopoulos, PF; Ceballos-Diaz, C; Golde, TE; Pratico, D. Adeno-associated virus-mediated brain delivery of 5-lipoxygenase modulates the AD-like phenotype of APP mice. *Mol Neurodegener.* 2012b, 7, 1. doi, 10.1186/1750-1326-7-1.

Chu, J; Lauretti, E; Di Meco, A; Praticò, D. FLAP pharmacological blockade modulates metabolism of endogenous tau *in vivo*. *Transl Psychiatry.* 2013a, 3, e333. doi, 10.1038/tp.2013.106.

Chu, J; Li, JG; Ceballos-Diaz, C; Golde, T; Praticò, D. The influence of 5-lipoxygenase on Alzheimer's disease-related tau pathology, *in vivo* and *in vitro* evidence. *Biol Psychiatry.* 2013b, 74, 321-8.

Chu, J; Li, JG; Praticò, D. Zileuton improves memory deficits; amyloid and tau pathology in a mouse model of Alzheimer's disease with plaques and tangles. *PLoS One.* 2013c, 8, e70991. doi, 10.1371/journal.pone.0070991.

Chu, J; Praticò, D. 5-lipoxygenase as an endogenous modulator of amyloid β formation *in vivo*. *Ann Neurol.* 2011a, 69, 34-46.

Chu, J; Praticò, D. 5-Lipoxygenase pharmacological blockade decreases tau phosphorylation *in vivo*, involvement of the cyclin-dependent kinase-5. *Neurobiol Aging.* 2013, 34, 1549-54.

Chu, J; Praticò, D. Involvement of 5-lipoxygenase activating protein in the amyloidotic phenotype of an Alzheimer's disease mouse model. *J Neuroinflammation.* 2012,;9, 127.

Chu, J; Praticò, D. Pharmacologic blockade of 5-lipoxygenase improves the amyloidotic phenotype of an Alzheimer's disease transgenic mouse model involvement of γ-secretase. *Am J Pathol.* 2011b, 178, 1762-9.

Connell, E; Darios, F; Broersen, K; Gatsby, N; Peak-Chew, SY; Rickman, C; Davletov, B. Mechanism of arachidonic acid action on syntaxin-Munc18. *EMBO Rep.* 2007, 8, 414-9.

Cui, JG; Kuroda, H; Chandrasekharan, NV; Pelaez, RP; Simmons, DL; Bazan, NG; Lukiw, WJ. Cyclooxygenase-3 gene expression in Alzheimer hippocampus and in stressed human neural cells. *Neurochem Res.* 2004, 29, 1731-7.

D'Addario, C; Di Francesco, A; Arosio, B; Gussago, C; Dell'Osso, B; Bari, M; Galimberti, D; Scarpini, E; Altamura, AC; Mari, D; Maccarrone, M. Epigenetic regulation of fatty acid amide hydrolase in Alzheimer disease. *PLoS One.* 2012, 7, e39186.

Darios, F; Ruipérez, V; López, I; Villanueva, J; Gutierrez, LM; Davletov, B. Alpha-synuclein sequesters arachidonic acid to modulate SNARE-mediated exocytosis. *EMBO Rep.* 2010, 11, 528-33.

Dawson, GR; Seabrook, GR; Zheng, H; Smith, DW; Graham, S; O'Dowd, G; Bowery, BJ; Boycen, S; Trumbauer, ME; Chen HY; Van der Ploeg, LH; Sirinathsinghji, DJ. Age-related cognitive deficits, impaired long-term potentiation and reduction in synaptic marker density in mice lacking the beta-amyloid precursor protein. *Neuroscience*. 1999; 90, 1-13.

Deane, R; Wu, Z; Zlokovic, BV. RAGE (yin) versus LRP (yang) balance regulates alzheimer amyloid beta-peptide clearance through transport across the blood-brain barrier. *Stroke*. 2004, 35, 2628-31.

Desbène, C; Malaplate-Armand, C; Youssef, I; Garcia, P; Stenger, C; Sauvée, M; Fischer, N; Rimet, D; Koziel, V; Escanyé, MC; Oster, T; Kriem, B; Yen, FT; Pillot, T; Olivier, JL. Critical role of $cPLA_2$ in Aβ oligomer-induced neurodegeneration and memory deficit. *Neurobiol Aging*. 2012, 33, 1123.e17-29.

Di Francesco, A; Arosio, B; Gussago, C; Dainese, E; Mari, D; D'Addario, C; Maccarrone, M. Involvement of 5-lipoxygenase in Alzheimer's disease, a role for DNA methylation. *J Alzheimers Dis*. 2013, 37, 3-8.

Dingwall, C. Spotlight on BACE, the secretases as targets for treatment in Alzheimer disease. *J Clin Invest*. 2001, 108, 1243-6.

Drouet, B; Pinçon-Raymond, M; Chambaz, J; Pillot, T. Molecular basis of Alzheimer's disease. *Cell Mol Life Sci*. 2000, 57, 705-15.

Emmerling, MR; Dudley, DT; Dyer, RD; Carroll, RT; Doyle, PD; Davis, RE. The role of arachidonic acid in the secretion of the amyloid precursor protein (APP). *Ann N Y Acad Sci*. 1996, 777, 310-5.

Emmerling, MR; Moore, CJ; Doyle, PD; Carroll, RT; Davis, RE. Phospholipase A2 activation influences the processing and secretion of the amyloid precursor protein. *Biochem Biophys Res Commun*. 1993, 197, 292-7.

Eriksen, JL; Sagi, SA; Smith, TE; Weggen, S; Das P; McLendon, DC; Ozols, VV; Jessing, KW; Zavitz, KH; Koo, EH; Golde, TE. NSAIDs and enantiomers of flurbiprofen target gamma-secretase and lower Abeta 42 *in vivo*. *J Clin Invest*. 2003, 112, 440-9.

Esposito, G; De Filippis, D; Carnuccio, R; Izzo, ARA; Iuvone, T. The marijuana component cannabidiol inhibits beta-amyloid-induced tau protein hyperphosphorylation through Wnt/beta-catenin pathway rescue in PC12 cells. *J Mol Med (Berl)*. 2006a, 84, 253-8.

Esposito, G; De Filippis, D; Maiuri, MC; De Stefano, D; Carnuccio, R; Iuvone, T. Cannabidiol inhibits inducible nitric oxide synthase protein expression and nitric oxide production in beta-amyloid stimulated PC12 neurons through p38 MAP kinase and NF-kappaB involvement. *Neurosci Lett.* 2006b, 399, 91-5.

Fabian, J; Hanekamp, W; Thomas, MH; Olivier, JL; Lehr, M. Investigations on the metabolic stability of cytosolic phospholipase $A_2\alpha$ inhibitors with 1-indolylpropan-2-one structure. *Chem Biol Interact.* 2013, 206, 356-63.

Fiebich, BL; Hüll, M; Lieb, K; Schumann, G; Berger, M; Bauer, J. Potential link between interleukin-6 and arachidonic acid metabolism in Alzheimer's disease. *J Neural Transm Suppl.* 1998, 54, 268-78.

Firuzi, O; Zhuo, J; Chinnici, CM; Wisniewski, T; Praticò, D. 5-Lipoxygenase gene disruption reduces amyloid-beta pathology in a mouse model of Alzheimer's disease. *FASEB J.* 2008, 22, 1169-78.

Fraser, SP; Suh, YH; Djamgoz, MB. Ionic effects of the Alzheimer's disease beta-amyloid precursor protein and its metabolic fragments. *Trends Neurosci.* 1997, 20, 67-72.

Fujimi, K; Noda, K; Sasaki, K; Wakisaka, Y; Tanizaki, Y; Iida, M; Kiyohara, Y; Kanba, S; Iwaki, T. Altered expression of COX-2 in subdivisions of the hippocampus during aging and in Alzheimer's disease, the Hisayama Study. *Dement Geriatr Cogn Disord.* 2007, 23, 423-31.

Fukaya, T; Gondaira, T; Kashiyae, Y; Kotani, S; Ishikura, Y; Fujikawa, S; Kiso, Y; Sakakibara, M. Arachidonic acid preserves hippocampal neuron membrane fluidity in senescent rats. *Neurobiol Aging.* 2007, 28, 1179-86.

Furukawa, K; Abe Y; Akaike, N. Amyloid beta protein-induced irreversible current in rat cortical neurones. *Neuroreport.* 1994, 5, 2016-8.

Gattaz, WF; Maras, A; Cairns, NJ; Levy, R; Förstl, H. Decreased phospholipase A2 activity in Alzheimer brains. *Biol Psychiatry.* 1995, 37, 13-7.

Giannopoulos, PF; Chu, J; Joshi, YB; Sperow, M; Li, JG; Kirby, LG; Praticò, D. Gene knockout of 5-lipoxygenase rescues synaptic dysfunction and improves memory in the triple-transgenic model of Alzheimer's disease. *Mol Psychiatry.* 2013a, doi, 10.1038/mp.2013.23.

Giannopoulos, PF; Chu, J; Joshi, YB; Sperow, M; Li, JG; Kirby, LG; Praticò, D. 5-lipoxygenase activating protein reduction ameliorates cognitive deficit; synaptic dysfunction; and neuropathology in a mouse model of Alzheimer's disease. *Biol Psychiatry.* 2013b, 74, 348-56.

Golej, DL; Askari, B; Kramer, F; Barnhart, S; Vivekanandan-Giri, A; Pennathur, S; Bornfeldt, KE. Long-chain acyl-CoA synthetase 4 modulates prostaglandin E_2 release from human arterial smooth muscle cells. *J Lipid Res.* 2011, 52, 782-93.

Goodman, Y; Steiner, MR; Steiner, SM; Mattson, MP. Nordihydroguaiaretic acid protects hippocampal neurons against amyloid beta-peptide toxicity; and attenuates free radical and calcium accumulation. *Brain Res.* 1994, 654, 171-6.

Haghani, M; Shabani, M; Javan, M; Motamedi, F; Janahmadi, M. CB1 cannabinoid receptor activation rescues amyloid β-induced alterations in behaviour and intrinsic electrophysiological properties of rat hippocampal CA1 pyramidal neurones. *Cell Physiol Biochem.* 2012, 29, 391-406.

Hansson, SF; Andréasson, U; Wall, M; Skoog, I; Andreasen, N; Wallin, A; Zetterberg, H; Blennow, K. Reduced levels of amyloid-beta-binding proteins in cerebrospinal fluid from Alzheimer's disease patients. *J Alzheimers Dis.* 2009, 16, 389-97.

Hawkes, CA; Shaw, JE; Brown, M; Sampson, AP; McLaurin, J; Carare, RO. MK886 Reduces Cerebral Amyloid Angiopathy Severity in TgCRND8 Mice. *Neurodegener Dis.* 2014, 13, 17-23.

Hicks, JB; Lai, Y; Sheng, W; Yang, X; Zhu, D; Sun, GY; Lee, JC. Amyloid-beta peptide induces temporal membrane biphasic changes in astrocytes through cytosolic phospholipase A2. *Biochim Biophys Acta.* 2008, 1778, 2512-9.

Hirabayashi, T; Murayama, T; Shimizu, T. Regulatory mechanism and physiological role of cytosolic phospholipase A2. *Biol Pharm Bull.* 2004, 27, 1168-73.

Ho, L; Purohit, D; Haroutunian, V; Luterman, JD; Willis, F; Naslund, J; Buxbaum, JD; Mohs, RC; Aisen, PS; Pasinetti, GM. Neuronal cyclooxygenase 2 expression in the hippocampal formation as a function of the clinical progression of Alzheimer disease. *Arch Neurol.* 2001, 58, 487-92.

Hoozemans, JJ; Veerhuis, R; Janssen, I; van Elk, EJ; Rozemuller ,AJ; Eikelenboom, P. The role of cyclo-oxygenase 1 and 2 activity in prostaglandin E secretion by cultured human adult microglia, implications for Alzheimer's disease. *Brain Res.* 2002, 951, 218-26.

Hsia, AY; Masliah, E; McConlogue, L; Yu, GQ; Tatsuno, G; Hu K; Kholodenko, D; Malenka, RC; Nicoll, RA; Mucke, L. Plaque-independent disruption of neural circuits in Alzheimer's disease mouse models. *Proc Natl Acad Sci U S A.* 1999, 96, 3228-33.

Ikonomovic, MD; Abrahamson, EE; Uz, T; Manev, H; Dekosky, ST. Increased 5-lipoxygenase immunoreactivity in the hippocampus of patients with Alzheimer's disease. *J Histochem Cytochem.* 2008, 56, 1065-73.

in 't Veld, BA; Launer, LJ; Hoes, AW; Ott, A; Hofman, A; Breteler, MM; Stricker, BH. NSAIDs and incident Alzheimer's disease. The Rotterdam Study. *Neurobiol Aging.* 1998, 19, 607-11.

Iwamoto, N; Kobayashi, K; Kosaka, K. The formation of prostaglandins in the postmortem cerebral cortex of Alzheimer-type dementia patients. *J Neurol.* 1989, 236, 80-4.

Jacobs, EH; Williams, RJ; Francis ,PT. Cyclin-dependent kinase 5; Munc18a and Munc18-interacting protein 1/X11alpha protein up-regulation in Alzheimer's disease. *Neuroscience.* 2006, 138, 511-22.

Janefjord, E; Mååg, JL; Harvey, BS; Smid, SD. Cannabinoid Effects on β Amyloid Fibril and Aggregate Formation; Neuronal and Microglial-Activated Neurotoxicity *In vitro. Cell Mol Neurobiol.* 2013, (*in press*)

Jin; M, Shepardson, N; Yang; T; Chen, G; Walsh, D; Selkoe, DJ. Soluble amyloid beta-protein dimers isolated from Alzheimer cortex directly induce Tau hyperphosphorylation and neuritic degeneration. *Proc Natl Acad Sci U S A.* 2011, 108, 5819-24.

Joshi, YB; Chu, J; Praticò, D. Knockout of 5-lipoxygenase prevents dexamethasone-induced tau pathology in 3xTg mice. *Aging Cell.* 2013a, 12, 706-11.

Joshi, YB; Di Meco, A; Praticó, D. Modulation of Amyloid-β Production by Leukotriene B4 via the γ-Secretase Pathway. *J Alzheimers Dis.* 2014, 38, 503-6.

Joshi, YB; Giannopoulos, PF; Chu, J; Praticò, D. Modulation of lipopolysaccharide-induced memory insult; γ-secretase; and neuroinflammation in triple transgenic mice by 5-lipoxygenase. Neurobiol Aging. 2013b, pii, S0197-458000584-8. doi, 10.1016/j.neurobiolaging. 2013b.11.016.

Jung, KM; Astarita, G; Yasar, S; Vasilevko, V; Cribbs, DH; Head, E; Cotman, CW; Piomelli, D. An amyloid β42-dependent deficit in anandamide mobilization is associated with cognitive dysfunction in Alzheimer's disease. *Neurobiol Aging.* 2012, 33, 1522-32.

Kagan, BL; Hirakura, Y; Azimov, R; Azimova, R; Lin, MC. The channel hypothesis of Alzheimer's disease, current status. *Peptides.* 2002, 23, 1311-5.

Kalifa, S; Polston, EK; Allard, JS; Manaye, KF. Distribution patterns of cannabinoid CB1 receptors in the hippocampus of APPswe/PS1ΔE9 double transgenic mice. *Brain Res.* 2011, 1376, 94-100.

Kanekiyo, T; Ban, T; Aritake, K; Huang, ZL; Qu, WM; Okazaki, I; Mohri, I; Murayama, S; Ozono, K; Taniike, M; Goto, Y; Urade, Y. Lipocalin-type prostaglandin D synthase/beta-trace is a major amyloid beta-chaperone in human cerebrospinal fluid. *Proc Natl Acad Sci U S A*. 2007, 104, 6412-7.

Kang, MJ; Fujino, T; Sasano, H; Minekura, H; Yabuki, N; Nagura, H; Iijima, H; Yamamoto, TT. A novel arachidonate-preferring acyl-CoA synthetase is present in steroidogenic cells of the rat adrenal; ovary; and testis. *Proc Natl Acad Sci U S A*. 1997, 94, 2880-4.

Kärkkäine, E; Tanila, H; Laitinen, JT. Functional autoradiography shows unaltered cannabinoid CB1 receptor signalling in hippocampus and cortex of APP/PS1 transgenic mice. *CNS Neurol Disord Drug Targets*. 2012, 11, 1038-44.

Kawamata, T; Taniguchi, T; Mukai, H; Kitagawa ,M; Hashimoto, T; Maeda, K; Ono, Y; Tanaka, C. A protein kinase; PKN; accumulates in Alzheimer neurofibrillary tangles and associated endoplasmic reticulum-derived vesicles and phosphorylates tau protein. *J Neurosci*. 1998, 18, 7402-10.

Kihara, T; Shimmyo, Y; Akaike, A; Niidome, T; Sugimoto, H. Abeta-induced BACE-1 cleaves N-terminal sequence of mPGES-2. Abeta-induced BACE-1 cleaves N-terminal sequence of mPGES-2. *Biochem Biophys Res Commun*. 2010, 393, 728-33.

Kimberly, WT; LaVoie, MJ; Ostaszewski, BL; Ye, W; Wolf,e MS; Selkoe, DJ. Gamma-secretase is a membrane protein complex comprised of presenilin; nicastrin; Aph-1; and Pen-2. *Proc Natl Acad Sci U S A*. 2001, 100, 6382-7.

Klyubin, I; Cullen, WK; Hu, NW; Rowan, MJ. Alzheimer's disease Aβ assemblies mediating rapid disruption of synaptic plasticity and memory. *Mol Brain*. 2012, 5, 25. doi, 10.1186/1756-6606-5-25.

Kochs, G; Hummel, R; Meyer, D; Hug, H; Marmé, D; Sarre, TF. Activation and substrate specificity of the human protein kinase C alpha and zeta isoenzymes. *Eur J Biochem*. 1993, 216, 597-606.

Köpke, E; Tung, YC; Shaikh, S; Alonso, AC; Iqbal, K; Grundke-Iqbal, I. Microtubule-associated protein tau. Abnormal phosphorylation of a non-paired helical filament pool in Alzheimer disease. *J Biol Chem*. 1993, 268, 24374-84.

Koppel, J; Bradshaw, H; Goldberg, TE; Khalili, H; Marambaud, P; Walker, MJ; Pazos, M; Gordon, ML; Christen, E; Davies, P. Endocannabinoids in Alzheimer's disease and their impact on normative cognitive performance, a case-control and cohort study. *Lipids Health Dis*. 2009, 8, 2.

Korte, M; Herrmann, U; Zhang, X; Draguhn, A. The role of APP and APLP for synaptic transmission, plasticity, and network function: lessons from genetic mouse models. *Exp Brain Res*. 2012, 217, 435-40.

Krafft, GA; Klein, WL. ADDLs and the signaling web that leads to Alzheimer's disease. *Neuropharmacology*. 2010, 59, 230-42.

Kriem, B; Sponne, I; Fifre, A; Malaplate-Armand, C; Lozac'h-Pillot, K; Koziel, V; Yen-Potin, FT; Bihain, B; Oster, T; Olivier, JL; Pillot, T. Cytosolic phospholipase A_2 mediates neuronal apoptosis induced by soluble oligomers of the amyloid-beta peptide. *FASEB J.* 2005, 19, 85-7.

Küch, EM; Vellaramkalayil, R; Zhang, I; Lehnen, D; Brügger, B; Stremmel, W; Ehehalt, R; Poppelreuther, M; Füllekrug, J. Differentially localized acyl-CoA synthetase 4 isoenzymes mediate the metabolic channeling of fatty acids towards phosphatidylinositol. *Biochim Biophys Acta.* 2013, 1841, 227-239.

Lacor, PN; Buniel, MC; Chang, L; Fernandez, SJ; Gong ,Y; Viola, KL; Lambert, MP; Velasco, PT; Bigio, EH; Finch, CE; Krafft, GA; Klein, WL. Synaptic targeting by Alzheimer's-related amyloid beta oligomers. *J Neurosci.* 2004, 24, 10191-200.

Larson, ME; Lesné, SE. Soluble Aβ oligomer production and toxicity. *J Neurochem.* 2012, 120 Suppl 1, 125-39.

Latham, CF; Osborne, SL; Cryle, MJ; Meunier, FA. Arachidonic acid potentiates exocytosis and allows neuronal SNARE complex to interact with Munc18a. *J Neurochem.* 2007, 100, 1543-54.

Laurén, J; Gimbel, DA; Nygaard, HB; Gilbert, JW; Strittmatter, SM. Cellular prion protein mediates impairment of synaptic plasticity by amyloid-beta oligomers. *Nature.* 2009, 457, 1128-32.

Lebeau, A; Esclaire, F; Rostène, W; Pélaprat, D. Baicalein protects cortical neurons from beta-amyloid [25-35] induced toxicity. *Neuroreport.* 2001, 12, 2199-202.

Lebeau, A; Terro, F; Rostene, W; Pelaprat, D. Blockade of 12-lipoxygenase expression protects cortical neurons from apoptosis induced by beta-amyloid peptide. *Cell Death Differ.* 2004, 11, 875-84.

Lee, HC; Inoue, T; Sasaki, J; Kubo T; Matsuda, S; Nakasaki, Y; Hattori, M; Tanaka, F; Udagawa, O; Kono, N; Itoh, T; Ogiso, H; Taguchi, R; Arita, M; Sasaki, T; Arai, H. LPIAT1 regulates arachidonic acid content in phosphatidylinositol and is required for cortical lamination in mice. *Mol Biol Cell.* 2012, 23, 4689-700.

Lee, JH; Agacinski, G; Williams, JH; Wilcock, GK; Esiri, MM; Francis, PT; Wong, PT; Chen, CP; Lai, MK. Intact cannabinoid CB1 receptors in the Alzheimer's disease cortex. *Neurochem Int.* 2010, 57, 985-9.

Liu, JW; Almaguel, FG; Bu, L; De Leon, DD; De Leon, M. Expression of E-FABP in PC12 cells increases neurite extension during differentiation, involvement of n-3 and n-6 fatty acids. *J Neurochem.* 2008, 106, 2015-29.

Liu, RZ; Mita, R; Beaulieu, M; Gao, Z; Godbout R. Fatty acid binding proteins in brain development and disease. *Int J Dev Biol.* 2010, 54[8-9]

Liu, Z; Huang, Y; Zhang, Y; Chen, D; Zhang, YQ. Drosophila Acyl-CoA synthetase long-chain family member 4 regulates axonal transport of synaptic vesicles and is required for synaptic development and transmission. *J Neurosci.* 2011, 31, 2052-63.

Liu, Z; Huang, Y; Zhang, Y; Chen, D; Zhang, YQ. Drosophila Acyl-CoA synthetase long-chain family member 4 regulates axonal transport of synaptic vesicles and is required for synaptic development and transmission. *J Neurosci.* 2011, 31, 2052-63.

Lovell, MA; Lynn, BC; Xiong, S; Quinn, JF; Kaye, J; Markesbery, WR. An aberrant protein complex in CSF as a biomarker of Alzheimer disease. *Neurology.* 2008, 70, 2212-8.

Luchsinger, JA; Cheng, D; Tang, MX; Schupf, N; Mayeux, R. Central obesity in the elderly is related to late-onset Alzheimer disease. *Alzheimer Dis Assoc Disord.* 2012, 26, 101-5.

Luchtman, DW; Song, C. Cognitive enhancement by omega-3 fatty acids from child-hood to old age, findings from animal and clinical studies. *Neuropharmacology.* 2013, 64, 550-65.

Lukiw, WJ. Gene expression profiling in fetal; aged; and Alzheimer hippocampus, a continuum of stress-related signaling. *Neurochem Res.* 2004, 29, 1287-97.

Lukiw, WJ; Bazan, NG. Cyclooxygenase 2 RNA message abundance; stability; and hypervariability in sporadic Alzheimer neocortex. *J Neurosci Res.* 1997, 50, 937-45.

Malaplate-Armand, C; Florent-Béchard, S; Youssef, I; Koziel, V; Sponne, I; Kriem, B; Leininger-Muller, B; Olivier, JL; Oster, T; Pillot, T. Soluble oligomers of amyloid-beta peptide induce neuronal apoptosis by activating a $cPLA_2$-dependent sphingomyelinase-ceramide pathway. *Neurobiol Dis.* 2006, 23, 178-89.

Martín-Moreno, AM; Brera, B; Spuch, C; Carro, E; García-García, L; Delgado, M; Pozo, MA; Innamorato, NG; Cuadrado, A; de Ceballos, ML. Prolonged oral cannabinoid administration prevents neuroinflammation; lowers β-amyloid levels and improves cognitive performance in Tg APP 2576 mice. *J Neuroinflammation.* 2012, 9, 8.

Maślińska, D; Kaliszek, A; Opertowska, J; Toborowicz, J; Deregowski, K; Szukiewicz, D. Constitutive expression of cyclooxygenase-2 (COX-2) in developing brain. A. Choroid plexus in human fetuses. *Folia Neuropathol.* 1999, 37, 287-91.

McLean, CA; Cherny, RA; Fraser, FW; Fuller, SJ; Smith, MJ; Beyreuther, K; Bush, AI; Masters, CL. Soluble pool of Abeta amyloid as a determinant of severity of neurodegeneration in Alzheimer's disease. *Ann Neurol.* 1999; 46(6), 860-6.

Meloni, I; Parri, V; De Filippis, R; Ariani, F; Artuso, R; Bruttini, M; Katzaki, E; Longo, I; Mari, F; Bellan, C; Dotti, CG; Renieri, A. The XLMR gene ACSL4 plays a role in dendritic spine architecture. *Neuroscience.* 2009, 159, 657-69.

Milton, NG. Anandamide and noladin ether prevent neurotoxicity of the human amyloid-beta peptide. *Neurosci Lett.* 2002, 332, 127-30.

Minghetti, L. Cyclooxygenase-2 [COX-2] in inflammatory and degenerative brain diseases. *J Neuropathol Exp Neurol.* 2004, 63, 901-10.

Miscia, S; Ciccocioppo, F; Lanuti, P; Velluto, L; Bascelli, A; Pierdomenico, L; Genovesi, D; Di Siena, A; Santavenere, E; Gambi, F; Ausili-Cèfaro, G; Grimley, PM; Marchisio, M; Gambi, D. Abeta(1-42) stimulated T cells express P-PKC-delta and P-PKC-zeta in Alzheimer disease. *Neurobiol Aging.* 2009, 30, 394-406.

Mita, R; Beaulieu, MJ; Field, C; Godbout, R. Brain fatty acid-binding protein and omega-3/omega-6 fatty acids, mechanistic insight into malignant glioma cell migration. *J Biol Chem.* 2010, 285, 37005-15.

Mohri, I; Kadoyama, K; Kanekiyo, T; Sato, Y; Kagitani-Shimono, K; Saito, Y; Suzuki, K; Kudo, T; Takeda, M; Urade, Y; Murayama, S; Taniike, M. Hematopoietic prostaglandin D synthase and DP1 receptor are selectively upregulated in microglia and astrocytes within senile plaques from human patients and in a mouse model of Alzheimer disease. *J Neuropathol Exp Neurol.* 2007, 66, 469-80.

Montine, KS; Quinn, JF; Zhang, J; Fessel, JP; Roberts, LJ 2nd; Morrow, JD; Montine, TJ. Isoprostanes and related products of lipid peroxidation in neurodegenerative diseases. *Chem Phys Lipids.* 2004, 128, 117-24.

Morgan, C; Colombres, M; Nuñez, MT; Inestrosa, NC. Structure and function of amyloid in Alzheimer's disease. *Prog Neurobiol.* 2004, 74, 323-49.

Morrow, JD; Awad, JA; Boss, HJ; Blair, IA; Roberts LJ 2nd. Non-cyclooxygenase-derived prostanoids [F2-isoprostanes] are formed in situ on phospholipids. *Proc Natl Acad Sci U S A.* 1992, 89, 10721-5.

Moses, GS; Jensen, MD; Lue, LF; Walker, DG; Sun, AY; Simonyi, A; Sun, GY. Secretory PLA2-IIA, a new inflammatory factor for Alzheimer's disease. *J Neuroinflammation.* 2006, 3, 28.

Mucke, L; Masliah, E; Yu, GQ; Mallory, M; Rockenstein, EM; Tatsuno, G; Hu, K; Kholodenko, D; Johnson-Wood, K; McConlogu,e L. High-level neuronal expression of abeta 1-42 in wild-type human amyloid protein precursor transgenic mice, synaptotoxicity without plaque formation. *J Neurosci.* 2000, 20, 4050-8.

Mukai, H. The structure and function of PKN; a protein kinase having a catalytic domain homologous to that of PKC. *J Biochem.* 2003, 133, 17-27.

Mulder, J; Zilberter, M; Pasquaré, SJ; Alpár, A; Schulte, G; Ferreira, SG; Köfalvi, A; Martín-Moreno, AM; Keimpema, E; Tanila, H; Watanabe, M; Mackie, K; Hortobágyi, T; de Ceballos, ML; Harkany ,T. Molecular reorganization of endocannabinoid signalling in Alzheimer's disease. *Brain.* 2011, 134, 1041-60.

Negre-Aminou, P; Pfenninger, KH. Arachidonic acid turnover and phospholipase A2 activity in neuronal growth cones. *J Neurochem.* 1993, 60, 1126-36.

Nicotra, A; Lupo, G; Giurdanella, G; Anfuso, CD; Ragusa N; Tirolo C; Marchetti B; Alberghina M. MAPKs mediate the activation of cytosolic phospholipase A2 by amyloid beta[25-35] peptide in bovine retina pericytes. *Biochim Biophys Acta.* 2005, 1733, 172-86.

Nikolaev, A; McLaughlin, T; O'Leary, DD; Tessier-Lavigne, M. APP binds DR6 to trigger axon pruning and neuron death via distinct caspases. *Nature.* 2009, 19, 457, 981-9.

Noonan, J; Tanveer, R; Klompas, A; Gowran, A; McKiernan, J; Campbell, VA. Endocannabinoids prevent β-amyloid-mediated lysosomal destabilization in cultured neurons. *J Biol Chem.* 2010, 285, 38543-54.

Novak, EM; Dyer, RA; Innis, SM. High dietary omega-6 fatty acids contribute to reduced docosahexaenoic acid in the developing brain and inhibit secondary neurite growth. *Brain Res.* 2008, 1237, 136-45.

Núñez, E; Benito, C; Tolón, RM; Hillard, CJ; Griffin, WS; Romero, J. Glial expression of cannabinoid CB receptors and fatty acid amide hydrolase are beta amyloid-linked events in Down's syndrome. *Neuroscience.* 2008, 151, 104-10.

Oka, S; Arita, H. Inflammatory factors stimulate expression of group II phospholipase A2 in rat cultured astrocytes. Two distinct pathways of the gene expression. *J Biol Chem.* 1991, 266, 9956-60.

Okaichi, Y; Ishikura, Y; Akimoto, K; Kawashima, H; Toyoda-Ono, Y; Kiso, Y; Okaichi, H. Arachidonic acid improves aged rats' spatial cognition. *Physiol Behav.* 2005, 84, 617-23.

Pamplona, FA; Ferreira, J; Menezes de Lima, O Jr; Duarte, FS; Bento, AF; Forner, S; Villarinho, JG; Bellocchio, L; Wotjak, CT; Lerner, R; Monory, K; Lutz, B; Canetti, C; Matias, I; Calixto ,JB; Marsicano, G; Guimarães, MZ; Takahashi, RN. Anti-inflammatory lipoxin A4 is an endogenous allosteric enhancer of CB1 cannabinoid receptor. *Proc Natl Acad Sci U S* A. 2012, 109, 21134-9.

Pan, B; Wang, W; Zhong, P; Blankman, JL; Cravatt, BF; Liu, QS. Alterations of endocannabinoid signaling; synaptic plasticity; learning; and memory in monoacylglycerol lipase knock-out mice. *J Neurosci.* 2011, 31, 13420-30.

Park, JH; Jung, MS; Kim, YS; Song, WJ; Chung, SH. Phosphorylation of Munc18-1 by Dyrk1A regulates its interaction with Syntaxin 1 and X11α. *J Neurochem.* 2012, 122, 1081-91.

Piccini, M; Vitelli, F; Bruttini, M; Pober, BR; Jonsson, JJ; Villanova, M; Zollo, M; Borsani, G; Ballabio, A; Renieri, A. FACL4; a new gene encoding long-chain acyl-CoA synthetase 4; is deleted in a family with Alport syndrome; elliptocytosis; and mental retardation. *Genomics.* 1998, 47, 350-8.

Pillot, T; Goethals, M; Vanloo, B; Talussot, C; Brasseur, R; Vandekerckhove, J; Rosseneu, M; Lins, L. Fusogenic properties of the C-terminal domain of the Alzheimer beta-amyloid peptide. *J Biol Chem.* 1996; 271, 28757-65.

Piro, JR; Benjamin, DI; Duerr, JM; Pi, Y; Gonzales, C; Wood, KM; Schwartz, JW; Nomura, DK; Samad, TA. A dysregulated endocannabinoid-eicosanoid network supports pathogenesis in a mouse model of Alzheimer's disease. *Cell Rep.* 2012, 1, 617-23.

Puccio, S; Chu, J; Praticò, D. Involvement of 5-lipoxygenase in the corticosteroid-dependent amyloid beta formation, *in vitro* and *in vivo* evidence. *PLoS One.* 2011, 6, e15163.

Reitz, C; Cheng, R; Rogaeva, E; Lee, JH; Tokuhiro, S; Zo,u F; Betten,s K; Sleegers, K; Tan, EK; Kimura, R; Shibata, N; Arai, H; Kamboh, MI; Prince, JA; Maier, W; Riemenschneider, M; Owen, M; Harold, D; Hollingworth, P; Cellini, E; Sorb,i S; Nacmia,s B; Takeda, M; Pericak-Vance, MA; Haines, JL; Younkin, S; Williams, J; van Broeckhoven, C; Farrer, LA; St George-Hyslop, PH; Mayeux, R. Genetic and Environmental Risk in Alzheimer Disease 1 Consortium. Meta-analysis of the association between variants in SORL1 and Alzheimer disease. *Arch Neurol.* 2011, 68, 99-106.

Rickman, C; Davletov, B. Arachidonic acid allows SNARE complex formation in the presence of Munc18. *Chem Biol.* 2005, 12, 545-53.

Rogaeva, E; Meng, Y; Lee, JH; Gu, Y; Kawarai, T; Zou, F; Katayama, T; Baldwin, CT; Cheng, R; Hasegawa, H; Chen, F; Shibata, N; Lunetta, KL; Pardossi-Piquard, R; Bohm, C; Wakutani, Y; Cupples, LA; Cuenco, KT; Green, RC; Pinessi, L; Rainer,o I; Sorbi, S; Bruni, A; Duar,a R; Friedland, RP; Inzelberg, R; Hampe, W; Bujo, H; Song, YQ; Andersen, OM; Willnow, TE; Graff-Radford, N; Petersen, RC; Dickson, D; Der, SD; Fraser, PE; Schmitt-Ulms, G; Younkin, S; Mayeu,x R; Farrer, LA; St George-Hyslop, P. The neuronal sortilin-related receptor SORL1 is genetically associated with Alzheimer disease. *Nat Genet.* 2007, 39, 168-77.

Rossner, S; Mehlhorn, G; Schliebs, R; Bigl, V. Increased neuronal and glial expression of protein kinase C isoforms in neocortex of transgenic Tg2576 mice with amyloid pathology. *Eur J Neurosci.* 2001, 13, 269-78.

Rowan, MJ; Klyubin, I; Wang, Q; Hu, NW; Anwyl R. Synaptic memory mechanisms, Alzheimer's disease amyloid beta-peptide-induced dysfunction. *Biochem Soc Trans.* 2007, 35(Pt 5), 1219-23.

Sagare, A; Deane, R; Bell, RD; Johnson, B; Hamm, K; Pendu, R; Marky, A; Lenting, PJ; Wu, Z; Zarcone, T; Goate, A; Mayo, K; Perlmutter, D; Coma, M; Zhong, Z; Zlokovic, BV. Clearance of amyloid-beta by circulating lipoprotein receptors. *Nat Med.* 2007, 13, 1029-31.

Sakurai, T; Kaneko, K; Okuno, M; Wada, K; Kashiyama, T; Shimizu, H; Akagi, T; Hashikawa, T; Nukina, N. Membrane microdomain switching, a regulatory mechanism of amyloid precursor protein processing. *J Cell Biol.* 2008, 183, 339-52.

Sanchez-Mejia, RO; Newman, JW; Toh, S; Yu, GQ; Zhou, Y; Halabisky, B; Cissé, M; Scearce-Levie, K; Cheng, IH; Gan, L; Palop, JJ; Bonventre, JV; Mucke, L. Phospholipase A_2 reduction ameliorates cognitive deficits in a mouse model of Alzheimer's disease. *Nat Neurosci.* 2008, 11, 1311-8.

Schaeffer, EL; De-Paula, VJ; da Silva, ER; de A Novaes, B; Skaf, HD; Forlenza, OV; Gattaz, WF. Inhibition of phospholipase A_2 in rat brain decreases the levels of total Tau protein. *J Neural Transm.* 2011, 118, 1273-9.

Schlosburg, JE; Blankman, JL; Long, JZ; Nomura, DK; Pan, B; Kinsey, SG; Nguyen, PT; Ramesh, D; Booker, L; Burston, JJ; Thomas, EA; Selley, DE; Sim-Selley, LJ; Liu, QS; Lichtman, AH; Cravatt, BF. Chronic monoacylglycerol lipase blockade causes functional antagonism of the endocannabinoid system. *Nat Neurosci.* 2010, 13, 1113-9.

Serrano-Pozo, A; Frosch, MP; Masliah, E; Hyman, BT. Neuropathological alterations in Alzheimer disease. *Cold Spring Harb Perspect Med.* 2011, 1, a006189.

Shaked, GM; Kummer, MP; Lu, DC; Galvan, V; Bredesen, DE; Koo, EH. Abeta induces cell death by direct interaction with its cognate extracellular domain on APP (APP 597-624). *FASEB J.* 2006, 20, 1254-6.

Shelat, PB; Chalimoniuk, M; Wang, JH; Strosznajder, JB; Lee, JC; Sun, AY; Simonyi ,A; Sun, GY. Amyloid beta peptide and NMDA induce ROS from NADPH oxidase and ARA release from cytosolic phospholipase A2 in cortical neurons. *J Neurochem.* 2008, 106, 45-55.

Solas, M; Francis, PT; Franco, R; Ramirez, MJ. CB2 receptor and amyloid pathology in frontal cortex of Alzheimer's disease patients. *Neurobiol Aging.* 2013, 34, 805-8.

Stephenson, DT; Manetta, JV; White, DL; Chiou, XG; Cox, L; Gitter, B; May, PC; Sharp, JD; Kramer, RM; Clemens, JA. Calcium-sensitive cytosolic phospholipase A2 (cPLA2) is expressed in human brain astrocytes. *Brain Res.* 1994, 637, 97-105.

Stephenson, D; Rash, K; Smalstig, B; Roberts, E; Johnstone, E; Sharp, J; Panetta, J; Little, S; Kramer, R; Clemens, J. Cytosolic phospholipase A_2 is induced in reactive glia following different forms of neurodegeneration. *Glia.* 1999, 27, 110-28.

Stephenson, DT; Lemere, CA; Selkoe, DJ; Clemens, JA. Cytosolic phospholipase A2 (cPLA2) immunoreactivity is elevated in Alzheimer's disease brain. *Neurobiol Dis.* 1996 Feb;3, 51-63.

Stumm, C; Hiebel, C; Hanstein, R; Purrio, M; Nagel, H; Conrad, A; Lutz, B; Behl, C; Clement, AB. Cannabinoid receptor 1 deficiency in a mouse model of Alzheimer's disease leads to enhanced cognitive impairment despite of a reduction in amyloid deposition. *Neurobiol Aging.* 2013, 34, 2574-84.

Sundaram, JR; Chan, ES; Poore, CP; Pareek, TK; Cheong, WF; Shui, G; Tang, N; Low, CM; Wenk, MR; Kesavapany, S. Cdk5/p25-induced cytosolic PLA_2-mediated lysophosphatidylcholine production regulates neuroinflammation and triggers neurodegeneration. *J Neurosci.* 2012, 32, 1020-34.

Szaingurten-Solodkin, I; Hadad, N; Levy, R. Regulatory role of cytosolic phospholipase A2alpha in NADPH oxidase activity and in inducible nitric oxide synthase induction by aggregated Abeta1-42 in microglia. *Glia.* 2009, 57, 1727-40.

Tanaka, T; Iwawaki, D; Sakamoto, M; Takai, Y; Morishige, J; Murakami, K; Satouchi, K. Mechanisms of accumulation of arachidonate in phosphatidylinositol in yellowtail. A comparative study of acylation systems of phospholipids in rat and the fish species Seriola quinqueradiata. *Eur J Biochem.* 2003, 270, 1466-73.

Tang, BL. Neuronal protein trafficking associated with Alzheimer disease, from APP and BACE1 to glutamate receptors. *Cell Adh Migr.* 2009, 3, 118-28.

Taniguchi, T; Kawamata, T; Mukai, H; Hasegawa, H; Isagawa, T; Yasuda M; Hashimoto T; Terashima A; Nakai M; Mori H; Ono Y; Tanaka C. Phosphorylation of tau is regulated by PKN. *J Biol Chem.* 2001, 276, 10025-31.

Tanveer, R; Gowran, A; Noonan, J; Keating, SE; Bowie, AG; Campbell, VA. The endocannabinoid; anandamide; augments Notch-1 signaling in cultured cortical neurons exposed to amyloid-β and in the cortex of aged rats. *J Biol Chem.* 2012, 287, 34709-21.

Thinakaran, G; Koo, EH. Amyloid precursor protein trafficking; processing; and function. *J Biol Chem.* 2008, 283, 29615-9.

Tomiyama, T; Matsuyama, S; Iso, H; Umeda, T; Takuma, H; Ohnishi, K; Ishibashi, K; Teraoka, R; Sakama, N; Yamashita, T; Nishitsuji, K; Ito, K; Shimada, H; Lambert, MP; Klein, WL; Mori H. A mouse model of amyloid beta oligomers, their contribution to synaptic alteration; abnormal tau phosphorylation; glial activation; and neuronal loss *in vivo. J Neurosci.* 2010, 30, 4845-56.

Tselepis, AD; John Chapman, M. Inflammation; bioactive lipids and atherosclerosis, potential roles of a lipoprotein-associated phospholipase A2; platelet activating factor-acetylhydrolase. *Atheroscler Suppl.* 2002, 3, 57-68.

Umeda, T; Tomiyama, T; Sakama, N; Tanaka, S; Lambert, MP; Klein, WL; Mori, H. Intraneuronal amyloid β oligomers cause cell death via endoplasmic reticulum stress; endosomal/lysosomal leakage; and mitochondrial dysfunction *in vivo. J Neurosci Res.* 2011, 89, 1031-42.

Valera, E; Dargusch, R; Maher, PA; Schubert, D. Modulation of 5-lipoxygenase in proteotoxicity and Alzheimer's disease. *J Neurosci.* 2013, 33, 10512-25.

Vetrivel, KS; Zhang, YW; Xu, H; Thinakaran, G. Pathological and physiological functions of presenilins. *Mol Neurodegener.* 2006,1, 4.

Vickers, PJ. 5-Lipoxygenase-activating protein (FLAP). *J Lipid Mediat Cell Signal.* 1995, 12, 185-94.

Vingtdeux, V; Marambaud, P. Identification and biology of α-secretase. *J Neurochem.* 2012, 120 Suppl 1, 34-45.

Walsh, DM; Selkoe, DJ. Deciphering the molecular basis of memory failure in Alzheimer's disease. *Neuron.* 2004, 44, 181-93.

Wang, DS; Dickson, DW; Malter, JS. beta-Amyloid degradation and Alzheimer's disease. *J Biomed Biotechnol.* 2006, 58406.

Wang, H; Ren, CH; Gunawardana, CG; Schmitt-Ulms, G. Overcoming barriers and thresholds - signaling of oligomeric Aβ through the prion protein to Fyn. *Mol Neurodegener.* 2013a, 8, 24.

Wang, J; Ueda, N. Biology of endocannabinoid synthesis system. *Prostaglandins Other Lipid Mediat.* 2009, 89, 112-9.

Wang, XY; Tang, SS; Hu, M; Long, Y; Li, YQ; Liao, MX; Ji, H; Hong, H. Leukotriene D4 induces amyloid-β generation via CysLTR-mediated NF-κB pathways in primary neurons. *Neurochem Int.* 2013b, 62, 340-7.

Westlake, TM; Howlett, AC; Bonner, TI; Matsuda, LA; Herkenham, M. Cannabinoid receptor binding and messenger RNA expression in human brain, an *in vitro* receptor autoradiography and in situ hybridization histochemistry study of normal aged and Alzheimer's brains. *Neuroscience.* 1994, 63, 637-52.

Willnow, TE; Carlo, AS; Rohe, M; Schmid,t V. SORLA/SORL1; a neuronal sorting receptor implicated in Alzheimer's disease. *Rev Neurosci.* 2010, 21, 315-29.

Wimo A; Prince M. World Alzheimer Report 2010. The Global Economic Impact of Dementia. *Alzheimer's Disease International.* Published by Alzheimer's Disease International (ADI) 21 September 2010. http://www.alz.co.uk/research/files/WorldAlzheimerReport2010.pdf

Wu, CC; Chawla, F; Game,s D; Ryde,l RE; Freedman, S; Schenk, D; Young, WG; Morrison JH; Bloom, FE. Selective vulnerability of dentate granule cells prior to amyloid deposition in PDAPP mice, digital morphometric analyses. *Proc Natl Acad Sci U S A.* 2004, 101, 7141-6.

Wu, J; Bie, B; Yang, H; Xu, JJ; Brown, DL; Naguib, M. Activation of the CB2 receptor system reverses amyloid-induced memory deficiency. *Neurobiol Aging.* 2013, 34, 791-804.

Yang, X; Sheng, W; He, Y; Cui, J; Haidekker MA; Sun GY; Lee JC. Secretory phospholipase A2 type III enhances alpha-secretase-dependent amyloid precursor protein processing through alterations in membrane fluidity. *J Lipid Res.* 2010, 51, 957-66.

Yasojima, K; Schwab, C; McGeer, EG; McGeer, PL. Distribution of cyclooxygenase-1 and cyclooxygenase-2 mRNAs and proteins in human brain and peripheral organs. *Brain Res.* 1999, 830, 226-36.

Yermakova, AV; Rollins, J; Callahan, LM; Rogers, J; O'Banion, MK. Cyclooxygenase-1 in human Alzheimer and control brain, quantitative analysis of expression by microglia and CA3 hippocampal neurons. *J Neuropathol Exp Neurol.* 1999, 58, 1135-46.

Yetimler, B; Ulusoy, G; Celik, T; Jakubowska-Doğru, E. Differential effect of age on the brain fatty acid levels and their correlation with animal cognitive status in mice. *Pharmacol Biochem Behav.* 2012, 103, 53-59.

Yu, W; Polepalli, J; Wagh, D; Rajadas, J; Malenka, R; Lu B. A critical role for the PAR-1/MARK-tau axis in mediating the toxic effects of Aβ on synapses and dendritic spines. *Hum Mol Genet.* 2012, 21, 1384-90.

Zach, S; Felk, S; Gillardon, F. Signal transduction protein array analysis links LRRK2 to Ste20 kinases and PKC zeta that modulate neuronal plasticity. *PLoS One.* 2010, 5, e13191.

Zemva, J; Schubert, M. The Role of Neuronal Insulin/ IGF-1 Signaling for the Pathogenesis of Alzheimer's Disease, *Possible Therapeutic Implications.* CNS Neurol Disord Drug Targets. 201. (*in press*)

Zhang, Y; Chen, D; Wang, Z. Analyses of mental dysfunction-related ACSl4 in Drosophila reveal its requirement for Dpp/BMP production and visual wiring in the brain. *Hum Mol Genet.* 2009, 18, 3894-905.

Zhu, D; Lai, Y; Shelat, PB; Hu, C; Sun, GY; Lee ,JC. Phospholipases A2 mediate amyloid-beta peptide-induced mitochondrial dysfunction. *J Neurosci.* 2006, 26, 11111-9.

In: Arachidonic Acid
Editor: Jason M. O'Keefe
ISBN: 978-1-63117-619-7

Chapter 4

ARACHIDONIC ACID, CYCLOOXYGENASE AND HEPATIC INFLAMMATION

Sura Wanessa Santos Rocha[1,2] and Christina Alves Peixoto[1]

[1]Ultrastructure Laboratory, Institute Aggeu Magalhães - FIOCRUZ, Pernambuco, Brazil
[2]Postgraduate Biological Sciences, Federal University of Pernambuco, Pernambuco, Brazil

ABSTRACT

Arachidonic acid is also subjected to reactions which leads to the formation of compounds known as leukotrienes. The first step in the formation of leukotrienes is cleavage of arachidonic acid from phospholipids membrane. The synthesis of leukotrienes begins with the action of several lipoxygenases, which catalyze the incorporation of molecular oxygen in arachidonic acid.

The arachidonic acid (AA) can also be converted by prostaglandin (PG) G/H synthase cytosolic, also known as cyclooxygenase (COX) in intermediates PGG2 and PGH2. The enzyme prostaglandin G/H synthase, has two sites catalyst: the site and the site cyclooxygenase peroxidase. The site cyclooxygenase converts the AA in PGG2, which in turn is reduced to the unstable intermediate, PGH2 by the peroxidase site, which is not inhibited by nonsteroidal anti-inflammatory drug (NSAIDS).

For NSAIDS treatment has been difficult to assign relative gastrointestinal "safety", since although they aimed COX-2 inhibition, the therapeutic doses of these drugs will also result in COX-1 reduced activity and severe hepatotoxicity. Therefore, there is a need to search for new therapeutic drugs used in the inhibition of COX-2 and other enzymes or inflammatory cytokines.

Some pharmacological studies showed that diethylcarbamazine (DEC) interferes with the arachidonic acid metabolism, acting as an anti-inflammatory drug. DEC is the drug most widely used in the treatment of lymphatic filariasis since 1947. Some studies performed on vertebrates show that this drug has several direct biochemical effects on different enzyme systems, including glycolysis, folate metabolism and activity of acetylcholinesterase.

DEC also has anti-inflammatory properties as a result of its interference with the arachidonic acid metabolism, which includes lipoxygenases (LOX) and cyclooxygenases (COX) enzymes.

For the first time, was demonstrated that DEC can decrease liver injury in experimental alcoholism and malnutrition model, and revealed a clinical potential of DEC for therapeutic anti-inflammatory and hepato-protective applications.

Keywords: Arachidonic Acid, Cyclooxygenase, Inflammation, Diethylcarbamazine, Anti-inflammatory, Liver

INTRODUCTION

Arachidonic acid (AA) performs a key regulatory role in cell physiology. It consists of fatty acid with 20 carbons which are released from membrane phospholipids by the enzyme phospholipase A2, which is activated by several stimuli (chemical, inflammatory, traumatic, mitogenic). All cells are capable of releasing arachidonic acid, but the products are highly enzyme specific depending on cell level [1]. In response to hormonal or other stimulus, such as phospholipase which is present in most mammalian cell types, the membrane phospholipids are hydrolyzed, releasing arachidonic acid [2]. The initial step of the synthesis of eicosanoids involves the cleavage of arachidonic acid from phospholipid cell membranes [3]. Phospholipase A_2 and phospholipase C and possibly diacilglicerolipase are involved in the release of arachidonic acid. Phospholipases are particularly relevant to the formation of eicosanoids, as the fatty acids must be in a free form to be used as substrates cyclooxygenase and lipoxygenase [4].

A group of enzymes present in the smooth endoplasmic converts arachidonic acid to prostaglandins, a process that begins with the formation of an endoperoxide, PGH_2 immediately followed by many other prostaglandins and thromboxanes [4]. The enzyme thromboxane synthase, in platelet form, converts PGH_2 in thromboxane A_2 from where other thromboxanes are derived.

Thromboxanes induce vessel constriction and blood platelet aggregation, the first step of blood coagulation [4]. Prostacyclin (PGI_2), the predominant product of the arachidonic acid through the action of cyclooxygenase, is formed from the vascular endothelium and subendothelium. This substance inhibits all recognized platelet aggregation agonists and also disaggregates previously aggregated platelets. PGI_2 inhibits the adhesion of platelets and neutrophils on the surfaces and injured endothelium, and dilates smooth muscles, such as the bronchial vascular muscle [1]. Arachidonic acid is also widely subjected to lipoxygenation reactions which lead to the formation of compounds known as leukotrienes. The first step in the formation of leukotrienes is the cleavage of arachidonic acid from the phospholipid membrane [5]. The synthesis of leukotrienes begins with the action of several lipoxygenases, which catalyze the incorporation of molecular oxygen in arachidonic acid. These enzymes are found in leukocytes in the heart, the brain, the lung and the spleen. The first is formed of metabolized leukotriene A_4, leukotriene B_4 (LTB_4) or leukotriene C_4 (LTC_4). The LTC_4 produces large-scale pulmonary bronchoconstriction, an effect that is partially blocked by cyclooxygenase inhibitors. This is able to mediate the effect of the LCT through the release of a prostaglandin bronchoconstrictor such as TxA_2 (thromboxane A_2). The LTB (leukotriene B_4) stimulates adhesion, migration, enzyme release and superoxide generation by polymorphonuclear leukocytes. These properties strongly biologically suggest a role for a hipoxigenase product such as inflammation antigen-induced bronchoconstriction [1].

There is an early organic inflammatory response when faced with tissue injury or infection. This physiological process involves a coordinated action between the system and the immune tissue in which the injury occurred [6].

Arachidonic acid (AA) can also be converted to prostaglandin (PG) G/H synthase cytosolic, also known as cyclooxygenase (COX) in intermediates PGG2 and PGH2.

The enzyme prostaglandin G/H synthase has two catalyst sites: the cyclooxygenase site and the peroxidase site. The cyclooxygenase site converts the AA into PGG2, which in turn is reduced into the unstable intermediate PGH 2 by the peroxidase site, which is not inhibited by NSAIDS.

The PGH2 is converted by tissue specific isomerasis multiple prostanoids (prostaglandins and thromboxanes). The cyclooxygenase enzyme has two isoforms entitled COX-1 and COX-2. Physiological studies have shown COX-1 constitutively acts as a cytoprotective gastric maintaining homeostasis and renal platelet COX-2 or inductive, which appears only in tissue trauma, inflammation and similar situations [2, 3]. The idea of specific COX-2 inhibitors would prevent process inflammation without undesirable side effects, particularly gastrointestinal disturbances, arising from the nonspecific blocking of COX. The term COX due to its proposed action consists of the formation of bicyclic peroxides (endoperoxides) from the oxygenation of polyunsaturated fatty acids [7]. In recent years, it has become clear that homologous isoforms of COX exist, known as COX-1 and COX-2 [8]. In almost all normal structural tissues the presence of COX-1 and low to undetectable levels of COX-2 were detected. These can be expressed in larger amounts by the presence of cytokines, growth factors and the stimulation of tumors, suggesting their relevance to cancerous and inflammatory processes [9]. The increased expression of COX-2 has also been implicated in diseases such as Alzheimer's and other neurological conditions. Thus, COX-1 was identified as constituent and COX-2 was described as inductive. The existence of a third isoform of this enzyme family, named COX-3 has recently been proposed. Unlike COX-1 and COX-2 this isoform would not produce pro-inflammatory prostanoids, but anti-inflammatory substances [10].

Such evidence could explain the periods of remission seen in some cases of chronic inflammatory diseases such as arthritis. The NSAIDS inhibit to varying degrees both COX isoforms in their therapeutic dosages. These have been characterized according to their ability to inhibit COX-1 and 2. Two agents that show some degree of COX-2 inhibition are meloxicam and nimesulide. It has been difficult to assign gastrointestinal "safety" to these compounds when compared with other conventional NSAIDs, since although COX-2 inhibition was preferred, the therapeutic doses of these drugs will also result in reduced COX-1 activity and severe hepatotoxicity. Prior to this there is a need to obtain new therapeutic drugs for use in the inhibition of COX-2 and other enzymes or inflammatory cytokines. Some pharmacological studies have shown that diethylcarbamazine (DEC) interferes with arachidonic acid metabolism, acting as an anti-inflammatory drug. Since 1947, DEC has been the most widely used drug in the treatment of lymphatic filariasis [11]. Some studies performed on vertebrates show that this drug has several direct biochemical effects on different enzyme systems, including glycolysis, folate metabolism and acetylcholinesterase activity [12].

DEC also has anti-inflammatory properties as a result of its interference with the arachidonic acid metabolism, which includes lipoxygenase (LOX) and cyclooxygenase (COX) enzymes [13, 14]. In studies carried out in association with the laboratory of the authors of the present study, Queto et al., (2010) demonstrated that DEC has an important role in blocking pulmonary eosinophilic inflammation in mice sensitized with ovalbumin, effectively preventing the effects of subsequent airway resistance, Th1/Th2 cytokine production, pulmonary eosinophil accumulation and eosinophilopoiesis in vivo and ex vivo.

Additionally, DEC directly suppressed IL-5–dependent eosinophilopoiesis in naive bone marrow cells [15]. DEC inhibits leukotriene synthesis by blocking the enzyme 5-lipoxygenase [13, 14]. It has also been demonstrated that DEC inhibits platelet aggregation, and thus has a possible effect on the cyclooxygenase (COX) pathway (Maizels 1992), which has similarities with the nitric oxide (NO) pathway, as both have constitutive and inducible isoforms of their enzymes and are key regulators of inflammatory responses [16, 17]. The studies of the authors analyzed the anti-inflammatory effect of DEC on hepatic cells of alcoholic mice. Thirty-two male C57BL/6 mice were equally divided into the following groups: a) a control group (C), which received only water, b) a DEC- treated group, which received 50 mg/Kg for 12 days (DEC50), c) and an alcoholic group (EtOH), submitted to only alcohol and d) an alcohol-DEC treated group (EtOH50), submitted to DEC treatment after the induction of chronic alcoholism for 6 weeks. Biochemical analyses were performed and liver fragments were processed for light microscopy, transmission electron microscopy, immunohistochemical and western blot. The level of AST increased significantly in the alcoholic group whereas a significant reduction of serum AST was detected in the EtOH50 group. Histological and ultrastructural analysis of the alcoholic group showed evident hepatocellular damage, which was considerably reduced in the alcoholic DEC-treated group. Immunohistochemistry results revealed a high expression of inflammatory markers such as IL-6, VCAM and ICAM by the hepatic cells of the EtOH group; however no immunoreactivity was detected for any of these cytokines after DEC treatment. Western blot analyses showed increased MCP-1 and iNOS expression in the EtOH group, which was significantly inhibited by DEC treatment.

According to the present results, DEC can be a potential drug for the treatment of chronic inflammation induced by chronic alcoholism [18]. Another study analyzed the DEC effect in the hepatocytes of malnourished mice.

Forty-eight male C57BL/6 mice were separated into groups: a control group (C, D25 and D50) and a malnourished group (M, MD25 and MD50).

After being induced to malnutrition, the mice were submitted to 12 days of treatment with orally administered DEC solutions in concentrations of 25 and 50 mg/kg. Biochemical analyses were performed and liver fragments were processed for light microscopy and transmission electron microscopy. For the enzymatic dosages, it was possible to observe a significant reduction in serum albumin in group M (2.52 ± 0.415 g/dL) when compared with the control group (C) (3.27 ± 0.427 g/dL) which received no treatment. Histological analysis of the M group showed evident hepatocellular damage. However, groups MD25 and MD50 returned to basal levels. Groups M, MD25 and MD50 presented a significant increase in alkaline phosphate when compared with the control group (C). Histological analysis of group M showed evident hepatic steatosis, which characterizes hepatocellular damage. Groups MD25 and MD50 showed a decrease in steatosis.

Ultrastructural analysis of the hepatocytes in groups MD25 and MD50 confirmed a reduction of lipid droplets. It is possible that the effects of DEC have contributed to the reduction of hepatic changes caused by malnutrition. For the first time, it was demonstrated that DEC decreases liver injury in experimental alcoholism and malnutrition models, and revealed the clinical potential of DEC for therapeutic anti-inflammatory and hepatoprotective applications [19].

REFERENCES

[1] Dahl, V., Aeder, J. C. - Non-opioid postoperative analgesia. *Acta Anaesthesiol. Scand.*, 2000;44: 1191-1203.

[2] Yaksh, T. L., Dirig, D. M., Conway, C. M., et al., - The acute antihyperalgesic action of nonsteroidal, anti-inflammatory drugs and release of spinal prostaglandin E2 is mediated by the inhibition of constitutive spinal cyclooxygenase-2 (COX-2) but not COX-1, *J. Neurosc.*, 2001;21:5847-5853.

[3] Perazella, M. A., Tray, K. - Selective cyclooxygenase-2 inhibitors: A pattern of nephrotoxicity similar to traditional nosteroidal anti-inflammatory drugs. *Am. J. Med.*, 2001;111:64-67.

[4] Souter, A. J., Fredman, B., White, P. F. - Controversies in the perioperative use of nonsteroidal antiinflamatory drugs. *Anesth. Analg.*, 1994;79:1178-1190.

[5] Tilley, S. L., Coffman, T. M., Koller, B. H. - Mixed messages modulation of inflammation and immune responses by prostaglandins and thromboxanes. *J. Clin. Invest.*, 2001;108:15-23.

[6] Marnett, L. J., Rowlinson, W. S., Goodwin, D. C., et al., - Arachidonic acid oxygenation by COX-1 and COX-2 – *J. Biol. Chem.*, 1999;274: 22903-22906.

[7] Jones, R. - Nonsteroidal Anti-inflamatory drug prescribing: past, present and future. *Am. J. Med.*, 2001;110:4s-7s.

[8] Fitzgerald, G. A., Patrono, C. - The coxibs, selective inhibitor of cyclooxygenase-2. *N. Engl. J. Méd.*, 2001; 345:433-442.

[9] Harris, R. C., Breyer, M. D. - Physiological regulation of cyclooxygenase- 2 in the kidney. *Am. J. Physiol. Renal Physiol.*, 2001; 281:F1-F11.

[10] Willoughby, D. A., Moore, A. R., Colville-Nash, P. R. - COX-1, COX-2, and COX-3 and the future of chronic inflammatory disease. *Lancet*, 2000;355:646-648.

[11] Hewitt, R. I. S., Kushner, H. W., Stewart, E., White, W. S., Wallace, Y. 1947. Experimental chemotherapy of filariasis III. Effect of 1-diethylcarbamazine-4-methylpiperazine hydrochloride against naturally acbquired filarial infections in cotton rats and dogs. *J. Lab. Clin. Med.* 32 (11): 1341-1329.

[12] Subrahmanyam, D. 1987. Antifilarials and their mode of action. *Ciba Encontradas Symp.* 127, 246-64.

[13] Maizels, R. M., Denham, D. A. 1992. Diethilcarbamazine (DEC): immunopharmacological interactions of an anti-filarial drug. *Parasitology* 105, 849-860.

[14] McGarry, H. L., Plant, L. D., Taylor, M. J. 2005. Diethylcarbamazine activity against Brugia malayi microfilariae is dependent on inducible nitric-oxide synthase and the cyclooxygenase pathway. *Filaria J.* 4, 4.

[15] Queto, T., Xavier, P., Gardel, M. A., Luca, B., et al., 2010. Inducible Nitric Oxide Synthase/CD95L-dependent Suppression of Pulmonary and Bone Marrow Eosinophilia by Diethylcarbamazine. *Am. J. Respir. Crit. Care Med.* 181, 429–437.

[16] Sano, H., Hla, T., Maier, J., Crofford, L., Case, J., Maciag, T., Wilder, R. L. 1992. In vivo cyclooxygenase expression in synovial tissues of patients with rheumatoid arthritis and osteoarthritis and rats with adjuvant and streptococcal cell wall arthritis. *J. Clin. Invest.* 89, 97-108.

[17] Clancy, R. M., Abramson, S. B. 1995. Nitric oxide: a novel mediator of inflammation. *Proc. Soc. Exp. Biol. Med.* 210 (2): 93-10.

[18] Rocha, S. W. S., Santos, A. C. O., Silva, B. S., Oliveira, D. C. T., Ribeiro, E. L., Barbosa, K. P. S., Gomes, F. O. S., Peixoto, C. A. Effects of Diethylcarbamazine (DEC) on hepatocytes of C57BL/6J mice submitted protein malnutrition. *JFDA*, 2012; 20(2): 524-531a.

[19] Rocha, S. W. S., Silva, B. S., Gomes, F. O. S., Silva, A. K. S., Raposo, C., Barbosa, K. P. S., Oliveira, D. C. T., Santos, A. C. O., Peixoto, C. A. Effect of diethylcarbamazine on chronic hepatic inflammation induced by alcohol in C57BL/6 mice. *Europ. J. Pharm.* 2012; 689(1-3):194-203b.

In: Arachidonic Acid
Editor: Jason M. O'Keefe

ISBN: 978-1-63117-619-7

Chapter 5

EFFECT OF ARACHIDONIC ACID AND ITS PRODUCING ENZYME PHOSPHOLIPASE $A_2\alpha$ ON KEY ONCOGENIC PATHWAYS IN PROSTATE CANCER

***Sheng Hua*[1]*, Chanlu Xie*[2]*, Mu Yao*[1] *and Qihan Dong*[1,2]**

[1]Discipline of Endocrinology and Central Clinical School and Bosch Institute, Royal Prince Alfred Hospital, The University of Sydney, Sydney, NSW, Australia

[2]School of Science and Health, The University of Western Sydney, Sydney, Australia

ABSTRACT

The dietary arachidonic acid (AA) is stored at the *sn-2* position in the glycerolphospholipid of cell membrane. AA is released *via* the cleavage by phospholipase A_2 in response to stimuli such as cytokines, growth factors and hormones. The freed AA is metabolised typically by cyclooxygenase and lipoxygenase enzymes to form eicosanoids. Of all the phospholipase A_2 that catalyse the hydrolysis of fatty acid at *sn-2* bond of glycerophospholipids, cytosolic phospholipase $A_2\alpha$ ($cPLA_2\alpha$) is preferentially targeting AA-containing phospholipids. Hence, $cPLA_2\alpha$ action represents the key step for AA release and eicosanoid synthesis. Stimulation of cells with agents that mobilise intracellular calcium and/or promote the phosphorylation of $cPLA_2\alpha$ leads to increased AA release and eicosanoid production.

AA and its metabolites have been shown to promote cancer cell proliferation and thus are under intensive investigation. Interestingly, the rate of eicosanoid production is also significantly higher in malignant than benign prostate tissues. AA increases the proliferation of prostate cancer cells *in vitro*. Mice on a diet with high AA content have more frequent and larger tumour recurrences following tumour removal. AA metabolites have been shown to affect cancer cell proliferation, invasion, angiogenesis and tumorigenesis. Blocking $cPLA_2\alpha$ inhibits proliferation of prostate cancer cells *in vitro* and *in vivo*. The mechanisms underlying AA stimulatory action on prostate cancer cell proliferation are likely due to an AA-mediated increase in activation of protein kinase B (AKT), extra-cellular regulated kinase (ERK) and androgen receptor (AR) signalling pathways.

1. INTRODUCTION

Arachidonic acid (AA) is a member of the ω-6 polyunsaturated fatty acids [1]. As an essential fatty acid, the dietary AA in blood is taken up and esterified at the *sn*-2 position in phospholipids of cell membranes [2, 3]. The membrane-bound AA is released following the cleavage by phospholipase A_2 [4, 5]. Freed AA is metabolised by cylooxygenase (COX) and lipoxygenase (LOX) enzymes to form eicosanoids [6]. As inflammation mediators [6-8], AA and its metabolites exert a wide range of actions including promoting prostate cancer cell proliferation [9]. Recent studies have begun to shed light on the effect of AA and phospholipase A_2 on key oncogenic pathways in prostate cancer cells and thus reveal the mechanism of promoting prostate cancer cell proliferation by AA and phospholipase A_2.

2. CYTOSOLIC PHOSPHOLIPASE A2α

Phospholipase A_2 family has over 20 identified members in mammals [10-12] and can be divided into different classes on the basis of their amino acid sequence homology. Many members are differentially expressed in a tissue-, species-, and genotype-specific manner [10, 13]. Of all the phospholipase A_2, cytosolic phospholipase $A_2\alpha$ ($cPLA_2\alpha$) plays a central role in the initiation of

AA metabolism because it preferentially targets AA-containing glycerophospholipids [14, 15].

Stimulation of cells with agents that mobilise intracellular calcium and/or promote the phosphorylation of $cPLA_2\alpha$ leads to: (i) translocation of the enzyme from the cytosol to the cell membrane, Golgi apparatus, endoplasmic reticulum and perinuclear membranes where it may associate with the glycerol phospholipid to access AA; and (ii) an increases in phospholipid binding affinity and AA release [16, 17].

2.1. $cPLA_2\alpha$ Structure

The *$cPLA_2\alpha$* coding region is located at chromosome 1q25 and is encoded by a 3.4 kb mRNA with 849 amino acids. Its promoter region contains binding sequence for transcriptional factors of Activator Protein 1, Protein Pea 3 and Nuclear Factor-Kappa B [18]. $cPLA_2\alpha$ is highly conserved, sharing over 95% amino acid identity between human and mouse. The orthologs in chickens, zebra fish and xenopus share over 80% amino acid identity with human. $cPLA_2\alpha$ protein structure consists of an N-terminal C2 domain and a C-terminal catalytic domain. The two domains are linked by a short and flexible peptide [19, 20] (Figure 1).

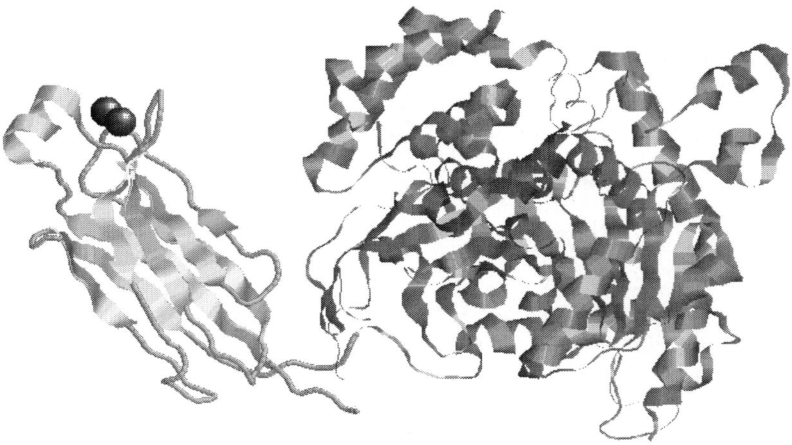

Figure 1. $cPLA_2\alpha$ structure in Rasmol. Ca^{2+} (red), C2 domain (yellow), catalytic domain (green) and a short and flexible peptide (purple).

C2 Domain

Interactions between cPLA$_2\alpha$ and membrane lipids occurs through the C2 domain, which consists of an eight-stranded antiparallel sandwich. Calcium binding to the C2 domain through Asp and Asn residues promotes association of hydrophobic residues with membrane phosphatidylcholine. This brings the catalytic subunit close to the membrane bilayer allowing enzymatic action to take place [19, 20].

Catalytic Domain

The cPLA$_2\alpha$ active site is a strictly conserved dyad composed of a serine (Ser^{228}) and an aspartic acid (Asp^{549}). A conserved Arginine (Arg^{200}) is also required for activity. The Ser/Asp active site of cPLA$_2\alpha$ lies in a deep funnel lined by hydrophobic residues, which confers the preference for AA-containing phospholipid [5]. The change in amino acid residues in the funnel decreases AA specificity and accounts for the change in sensitivity to some cPLA$_2\alpha$ inhibitors [21, 22].

In addition to the C2 domain, cPLA$_2\alpha$ also contains several conserved serine residues as phosphorylation sites for further regulation. Ser^{505} is the most studied and recognised site for phosphorylation, followed by Ser^{515} and Ser^{727}. Although phosphorylation is not necessary for basal enzyme activity, phosphorylation at each of these sites has shown to augment AA release to a different degree [23-25]. The kinase responsible for the phosphorylation at Ser^{505}, Ser^{515}, and Ser^{727} is p38 mitogens activated protein kinase, calcium/calmodulin-dependent protein kinase, and mitogen-activated protein interacting kinases, respectively [23, 24, 26, 27], suggesting differential regulation of cPLA$_2\alpha$ *via* different signalling processes.

2.2. cPLA$_2\alpha$ Function

In response to increases in cytokines, growth factors and Ca^{2+}, cPLA$_2\alpha$ recognises the *sn*-2 acyl bond of phospholipids and catalyses the hydrolysis to release the fatty acid and a lysophospholipid [28, 29]. The freed AA and lysophospholipids contribute to the production of eicosanoids and platelet activating factors, respectively. cPLA$_2\alpha$ also possesses lysophospholipase and transacylase activities but these activities are relatively low compared to AA-producing activity [4]. As cPLA$_2\alpha$ is the only phospholipase A_2 that displays a high preference for AA, it has been thought to be the rate-limiting step in eicosanoid synthesis [28]. The freed AA is rapidly oxidised by COX, LOX and P450-dependent epoxygenase to produce prostaglandins (PG), hydroxyeicosatetraenoic acids (HETE) and epoxyeicosatrienoic epoxides (EET) respectively. These eicosanoids regulate inflammation and promote mitogenesis and angiogenesis by binding to G-protein-coupled receptors (GPCRs). The other product of $cPLA_2\alpha$ action, lysophospholipid, is the precursor of platelet-activating factor and functions as a potent inflammatory mediator *via* GPCRs [30]. Thus, targeting of $cPLA_2\alpha$ offers an attractive option in suppressing both AA and lysophospholipid pathways.

3. Arachidonic Acid, $cPLA_2\alpha$ and Prostate Cancer

Prostate cancer (PC) is one of the most common cancers and the second most common cause of cancer-related mortality exceeded only by lung cancer [31, 32]. The incidence of PC varies 50-fold world-wide [33] and is most common in Western countries and least common in South-east Asia [34]. While prostate-confined cancer may be treated with prostatectomy and irradiation, androgen ablation therapy by surgical or medical castration becomes the mainstay of treatment options for metastasised PC [35]. Castration induces apoptosis in the majority of PC cells, which translates clinically to improvement in cancer-related symptoms and lowering of AR-regulated genes, including prostate specific antigen (PSA). Although PC is initially androgen dependent or responsive to castration, over time and/or during the course of treatment, it progresses to become castration resistant or hormone refractory.

AA and Eicosanoids

Recent studies have shown that dietary lipid intake contributes to PC progression. High AA intake reduces the time required to convert hormone sensitive to hormone refractory PC [36]. Mice supplemented with AA in the diet show earlier, more frequent, and larger tumor recurrence than controls after surgical removal of PC xenograft mimicking prostatectomy [37]. Also, dietary AA increases the tumor growth in prostate-specific phosphatase and tensin homolog (PTEN)-knockout mice [38]. Treatment of human PC cell line PC-3 with AA increases cell proliferation [39] and induces steroid synthesis contributing to androgen receptor signalling in PC cell line LNCaP [40]. These studies demonstrate a role for AA in prostate cancer progression or recurrence.

In addition, *ex vivo* study shows that the eicosanoid synthesis rates from the radiolabeled AA are higher in cancerous than in benign prostate tissue, suggesting an increased utilisation of AA in prostate cancer [41]. In two clinical trials, the rate of PSA increase after prostatectomy or radiotherapy is slowed in patients treated with inhibitors of eicosanoid-producing enzymes [42, 43]. These studies demonstrate that eicosanoids play a role in prostate cancer progression or recurrence.

PLA_2

We have shown an increase in the secretory PLA_2 [44] and phospho-$cPLA_2\alpha$ [45] protein levels and a diminished Annexin II protein level [46, 47] in human prostate cancer. In patients receiving androgen ablation therapy, secretory PLA_2-IIA (sPLA2-IIA) remains elevated in surviving cancer cells but is substantially decreased in benign glands. Exogenous addition of $sPLA_2$-IIA, but not an inactive $sPLA_2$-IIA mutant (H48Q), results in a dose-dependent increase in cell proliferation, whereas $sPLA_2$-IIA–selective inhibitors suppress the cell proliferation in a dose dependent manner. As the effect of exogenous $sPLA_2$-IIA can be blocked by inhibition of $cPLA_2\alpha$, it suggests a role for $cPLA_2\alpha$ in mediating $sPLA_2$-IIA action. Considering the inhibitory effect on cell proliferation by $sPLA_2$ blockade appears to be *via* $cPLA_2\alpha$ [44], we have focused our study on $cPLA_2\alpha$ herein.

$cPLA_2\alpha$ is present in all prostate cancer cells lines. Inhibition of $cPLA_2\alpha$ by small interfering RNA or inhibitor with a concomitant reduction in $cPLA_2\alpha$ activity results in significant reductions in prostate cancer cell numbers, as a result of reduced proliferation as well as increased apoptosis. PC-3 xenograft growth is reduced by 33% *in vivo* [45]. Hence, $cPLA_2\alpha$ serves as a potentially effective therapeutic target for castration resistant PC.

4. Arachidonic Acid, $cPLA_2\alpha$ and Oncogenic Pathway PI3K/AKT in Prostate Cancer

4.1. The PI3K/AKT Pathway Overview

The phosphoinositide-3-kinase (PI3K) and downstream protein kinase B (AKT) regulate diverse cellular processes [48]. This PI3K/AKT pathway is highly conserved and tightly controlled *via* a multistep process [49]. PI3K/AKT kinases act as major switches conveying signals from transmembrane receptor tyrosine kinases (RTKs) [48]. Activation of PI3K is initiated by extracellular stimuli through binding to RTK. These stimuli include growth-related factors such as epidermal growth factor (EGF), platelet-derived growth factor (PDGF), vascular endothelial growth factor (VEGF) or metabolism-related factors such as insulin and insulin-like growth factor (IGF) [50]. Ligand binding to the RTK monomer leads to receptor dimerisation, phosphorylation and activation. This activated form of RTK is then able to activate PI3K directly or through recruitment of adaptor molecules, such as insulin receptor substrate protein (IRS) or growth factor receptor-bound protein 2 (Grb-2) [51].

PI3K is a heterodimeric protein with two subunits: p85 (regulatory subunit) and p110 (catalytic subunit). Several isoforms of both subunits exit, although more for p110 than p85 [48]. The p85 subunit inhibits and stabilises the p110 subunit until binding to RTK through its Src-Homology 2 (SH2) domain [52] causing phosphorylation at Tyr and Thr residues of the p85 subunit, which leads to a conformational change and thereby releases the inhibitory effect on the p110 subunit [53, 54]. As a lipid-modifying enzyme, the catalytic p110 subunit phosphorylates phosphatidylinositol-(4,5)-diphosphate (PIP_2) to phosphatidylinositol-(3,4,5)-triphosphate (PIP_3) [55]. The PIP_3 provides a docking site for AKT kinase to stay on the membrane through binding to its pleckstrin homology (PH) domain [49]. PIP3 the docking site allows AKT to be brought in close proximity to other regulatory molecules that phosphorylate (activate) or dephosphorylate (deactivate) AKT. In converse, PTEN reverse PIP3 back to PIP2 though dephosphorylation [56]. Thus, PI3K and PTEN is positive and negative regulator of AKT activity, respectively [56].

The AKT kinase phosphorylates a wide range of downstream targets, leading to increased proliferation, survival, and growth. AKT also influences the cellular response to nutrient availability, intermediary metabolism, angiogenesis and metastasis [57].

4.2. The PI3K/AKT Pathway in Prostate Cancer

The PI3K/AKT pathway is one of the most prevalent pathways deregulated in cancer including PC [48, 55]. Genetic alterations of PI3K and PTEN have been found in at least sixteen types of human cancers [48]. Studies have shown over 43% of prostate cancers have either a gain of function mutation in *PI3K* (the oncogene) or a loss of function mutation in *PTEN* (tumor suppressor) [48].

Although the PI3K/AKT pathway represents an important therapeutic target for treatment of PC, the complexities in signalling regulation pose substantial challenges in therapeutic design [51]. While inhibition of downstream effectors of AKT (e.g., blocking mTORC with rapamycin or RAD001) may seem effective, the process is complicated by a negative feedback regulation of upstream AKT [58]. mTOR inhibition inactivates its targeting S6 kinase, which serves as a negative regulator of PI3K through phosphorylation and inhibition of insulin receptor substrate-1 [59]. A decrease in this negative regulator results in increased AKT activation [51]. Hence, understanding the regulation of AKT signalling pathway is of high importance.

4.3. AA and Eicosanoids Promote AKT Signalling

AA and eicosanoid have been frequently implicated as pro-proliferative signals in cancer [60-63]. AA is able to increase prostate cancer cell proliferation and PI3K/AKT signalling in prostate cancer cells [39, 64]. AA metabolites such as PGE2, PGF2α, 5-HETE and 12-HETE have all been implicated in activating AKT in various cancer types [60-63]. Prostanoid receptors EP1, EP4 and FP have been shown to convey the eicosanoid signals to AKT pathways *via* transactivation of EGFR [62, 65]. Blocking eicosanoid production with non-steroidal anti-inflammatory drugs decreases the levels of phospho-AKT (pAKT) [66] and phospho-PI3K p85 subunit [64].

The tumor suppressor PTEN is often inactivated in prostate cancers causing an increase in pAKT. This may be due to mutation or truncation of PTEN as shown in prostate cancer cell lines LNCaP or PC-3 respectively [67, 68] or due to the effect of reactive oxygen species (ROS) [69]. AA metabolism by COX or LOX enzymes produces ROS, which is able to inactivate PTEN leading to increased AKT signalling [69]. Mechanistically, ROS induces oxidation on Cys^{124} of the PTEN catalytic domain and the resulting intramolecular disulfide bond inactivates PTEN function [70]. As COX and LOX are often overexpressed in inflammation and during multistage tumor progression [41, 71], the increased ROS synthesis is likely to contribute to the activation of AKT pathway in cancer.

4.4. Effect of Increase or Decrease in cPLA$_2$α Action on AKT Signalling

Very few studies have examined the effect of cPLA$_2$α action directly on prostate cancer. Following our finding that phospho-cPLA$_2$α is aberrantly elevated in castration resistant PC and pharmacological blocking of cPLA$_2$α can lead to a reduction of PC cell proliferation, we recently determined the effect of overexpression of cPLA$_2$α on AKT pathway in two prostate cancer cell lines with PTEN deletion or mutation. Indeed, overexpression of cPLA$_2$α with a resultant increase in phospho-cPLA$_2$α leads to an increase of pAKT and its immediate downstream effector pGSK3β and cyclin D1 [72]. In contrast, pharmacological blockade of cPLA$_2$α with Efipladib is able to attenuate the pAKT increase induced by cPLA$_2$α overexpression [72].

cPLA$_2$α silencing through the use of siRNA reduces AKT phosphorylation and its signalling in PC-3 and LNCaP cells (Figure 2). The same result can be achieved using a pharmacological agent blocking cPLA$_2$α (Figure 3). Further, in EGF-stimulated cells, the rise in AKT activation and signalling is abrogated by cPLA$_2$α silencing.

Since these findings were obtained from cell lines that are PTEN-null, it demonstrates the requirement of cPLA$_2$α in the constitutively activated AKT pathway in prostate cancer cells. To demonstrate that the effect of cPLA$_2$α on AKT is through AA, a treatment of prostate cancer cell lines (PC-3 and LNCaP) with AA reverts the reduction in pAKT brought about by cPLA$_2$α siRNA (Figure 4). These findings indicate the significance of cPLA$_2$α in maintaining AKT signalling in PC.

To determine the importance of eicosanoid in mediating AA action, we firstly determined the HETE levels following incubation of prostate cancer cells with AA. The 5-LOX product 5-HETE was amongst one of the highest HETEs [72]. The 5-LOX activity is critical in regulation of cancer cell proliferation [73, 74]. Inhibition of 5-LOX activity abrogates 5-LOX's pro-proliferative effects [66].

We then demonstrated that the action of cPLA$_2$α and AA in sustaining pAKT is *via* at least partly 5-HETE, as the AA stimulatory effect is nullified when 5-LOX activity is inhibited with MK886 (Figure 4). Hence, the effect of cPLA$_2$α action is likely mediated through AA and eicosanoids including 5-HETE.

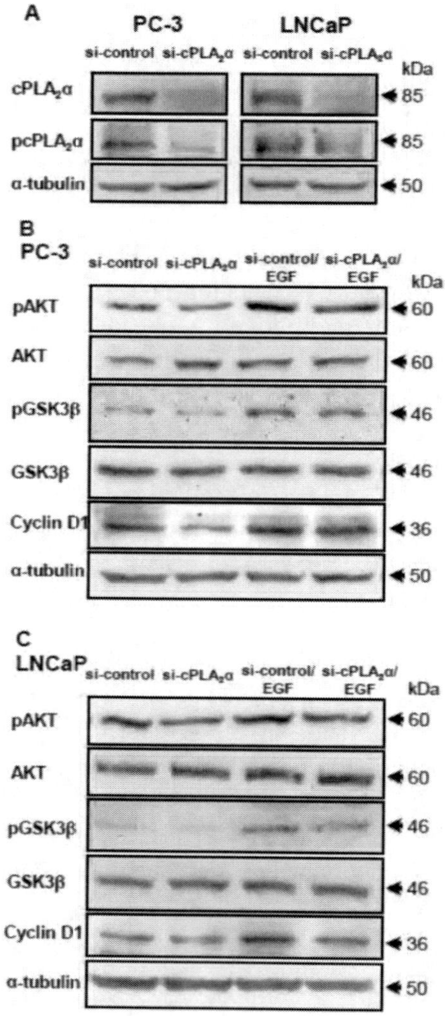

Figure 2. cPLA$_2$α inhibition by siRNA in LNCaP and PC3 cells lead to a decrease in pAKT and its signalling under basal or EGF stimulated conditions [72]. Genetic silencing of cPLA$_2$α diminishes AKT signalling by immunoblot. (A) A decrease in total and phospho-cPLA$_2$α in PC-3 and LNCaP cells following a 48 h/37 °C treatment with siRNA against cPLA$_2$α (si-cPLA$_2$α) relative to the scrambled control (si-control). The marked reduction of bands at 85 kDa by cPLA$_2$α siRNA also verified the bands detected by anti-total and phosphorylated cPLA$_2$α antibodies are cPLA$_2$α. (B) PC-3 cells and (C) LNCaP cells show reduced pAKT, pGSK3B and cyclin D1 levels under basal or EGF-stimulating (20 ng/mL/30 min) conditions.

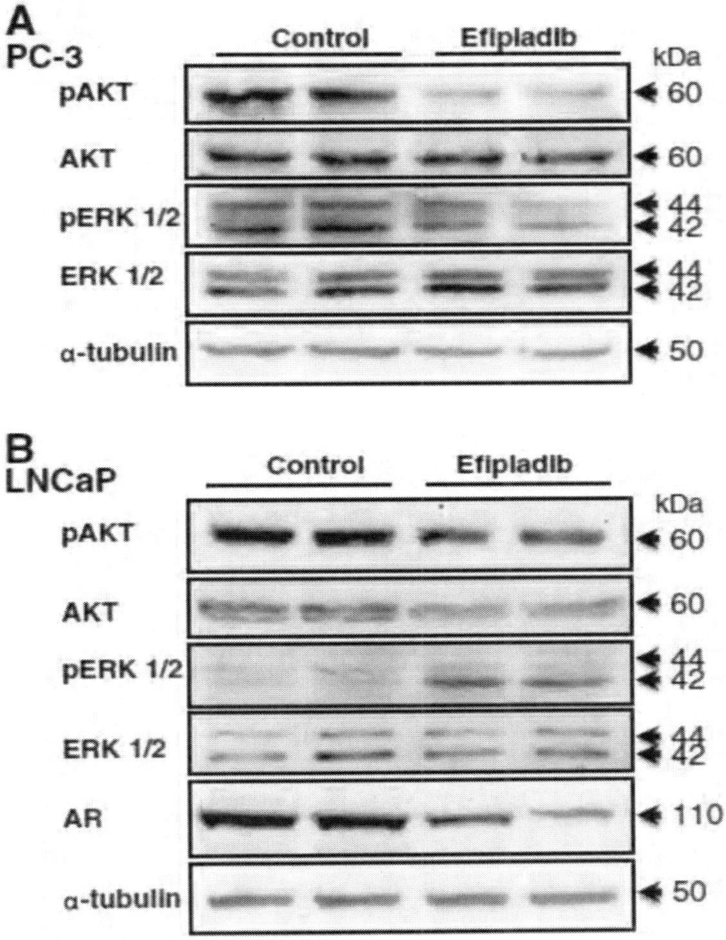

Figure 3. Effects of inhibition of $cPLA_2\alpha$ with Efipladib on AKT, ERK 1/2 and AR levels and cellular proliferation [72]. PC-3 and LNCaP cells were incubated with Efipladib (20 μM/72 h/37 °C) relative to the vehicle control (0.04% DMSO). (A) Immunoblot of pAKT and pERK 1/2 levels in PC-3 cells. (B) Immunoblot of pAKT, pERK 1/2 and AR levels in LNCaP cells.

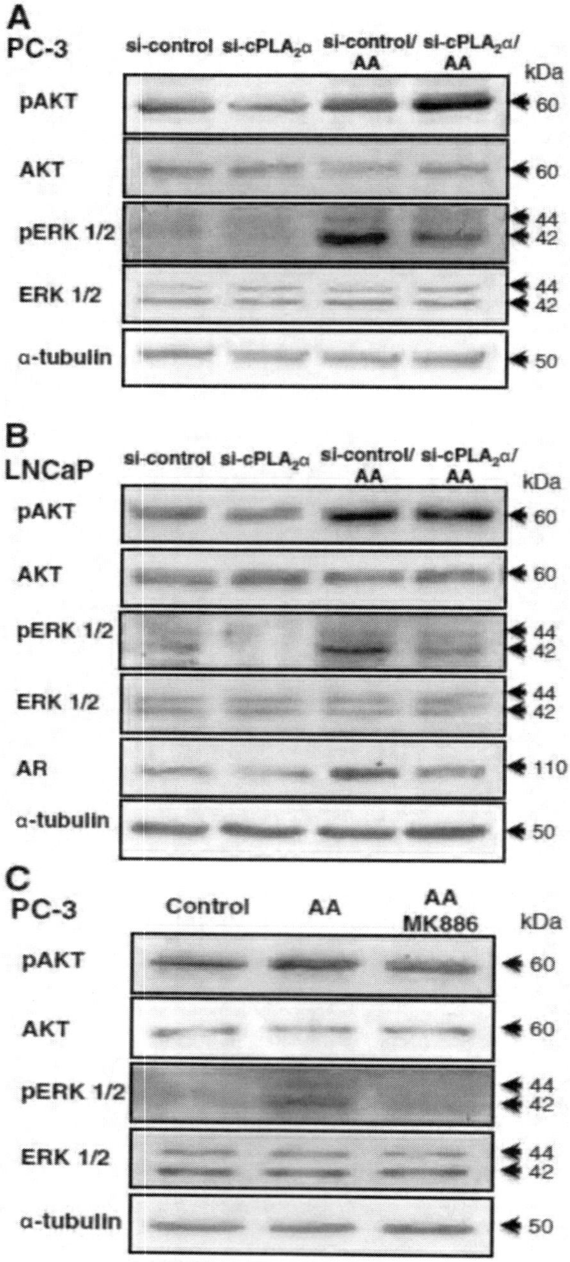
A
PC-3
si-control
si-cPLA2α
si-control/ AA
si-cPLA2α/ AA
kDa
pAKT
60
AKT
60
pERK 1/2
44
42
ERK 1/2
44
42
α-tubulin
50
B
LNCaP
si-control
si-cPLA2α
si-control/ AA
si-cPLA2α/ AA
kDa
pAKT
60
AKT
60
pERK 1/2
44
42
ERK 1/2
44
42
AR
110
α-tubulin
50
C
PC-3
Control
AA
AA MK886
kDa
pAKT
60
AKT
60
pERK 1/2
44
42
ERK 1/2
44
42
α-tubulin
50

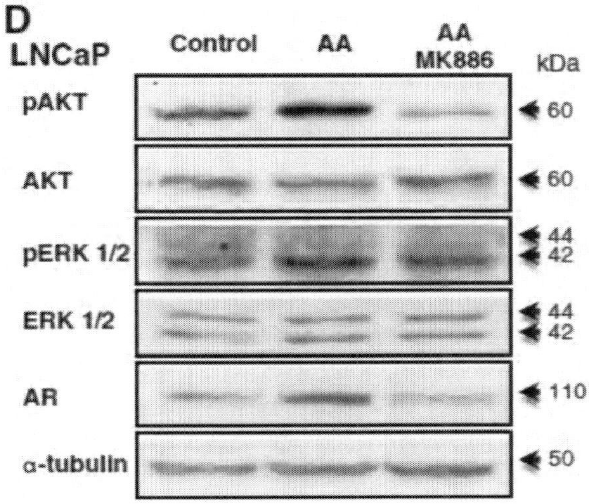

Figure 4. Effects of AA on AKT, ERK 1/2 and AR levels as determined by immunoblot [72]. (A) PC-3 and (B) LNCaP cells were treated with or without AA (20 μM) for 4 h/37 °C following a 48 h/37 °C incubation with si-cPLA$_2$α relative to the scrambled control. (C) PC-3 and (D) LNCaP cells treated with AA (20 μM/4 h/37 °C) in the presence of the 5-LOX inhibitor, MK886, show the diminished effect of AA alone on pAKT, pERK 1/2 and AR levels.

5. Archidonic Acid, cPLA$_2$α and ERK Pathway in Prostate Cancer

5.1. The ERK Pathway Overview

The RAF/MEK/ERK pathway is a part of the MAPK pathway and is conserved in all mammals [75, 76]. ERK has been established as a major component in the regulation of cell cycle progression, differentiation and apoptosis [77]. When improperly activated, ERK contributes to malignant transformation. The RAF/MEK/ERK pathway is regulated by kinases, transcription factors, apoptotic regulators and caspase activators [78]. Abnormal activation of ERK occurs in human cancer due to mutations in RAS or RAF or genes in other pathways (e.g., PI3K/AKT) that can influence ERK activity. Nearly 50% of human cancers have alterations in the RAS/RAF/MEK/ERK pathway or in components of RAS effector pathways that may lead to increased ERK activity [79]. The pathway also influences chemotherapeutic drug resistance as ectopic activation of RAF induces drug resistance in breast cancer cells [78]. Taken together, the ERK is an important MAPK for therapeutic targeting in cancer treatment.

The RAS/RAF/MEK/ERK signalling cascade is activated by a wide variety of receptors involved in growth and differentiation. The architecture of the pathway usually includes a set of adaptor proteins linking the receptor to a guanine nucleotide exchange factor (GEF) and then transducing the signal to the small GTP-binding RAS [80]. For instance, EGFR, activated by EGF binding, recruits the adaptor Grb2, which recruits the protein SOS. The docking of the Grb2-SOS complex on EGFR activates SOS and allows it to catalyse the replace of GDP on RAS with a GTP [81]. RAS-GTP is the active form of RAS and bind to RAF to activate RAF kinase activity [82]. Activated RAF kinase phosphorylates and activates MEK1 and MEK2, which in turn phosphorylate and activate ERK1 and ERK2 [78]. Activated ERK has over a hundred known targets [78] including transcription factors such as Ets-1, c-Jun and c-Myc [78], which are important for the cell cycle regulation.

5.2. The ERK Pathway in Prostate Cancer

Increased activation of the RAS/RAF/MEK/ERK pathway has been associated with advanced PC, hormonal resistance and a poor prognosis [83-87]. Additionally, RAS/RAF/MEK/ERK signalling is also implicated in PC progression by making androgen receptor hypersensitive to androgens [85]. There are evidences of cross-talk between PI3K/AKT and the RAS/RAF/MEK/ERK pathway. An increase in pAKT together with a decrease in phospho-ERK (pERK) in poorly differentiated PC has been reported [88-90]. This is in agreement with the notion that AKT can down-regulate RAF [78, 91] and indicates an intricacy in a therapeutic intervention of either AKT or ERK pathway.

5.3. cPLA$_2$α, Arachidonic Acid and Eicosanoids Promote ERK Signalling

AA metabolites such as PGE2, PGF2α, 5-HETE and 12-HETE have been shown to activate pERK in various cancer types [60-63]. Prostanoid receptors convey the eicosanoid signals to ERK pathways *via* transactivation of EGFR

[62, 65]. Blocking of eicosanoid production by non-steroidal anti-inflammatory drugs decreases the levels of pERK [65].

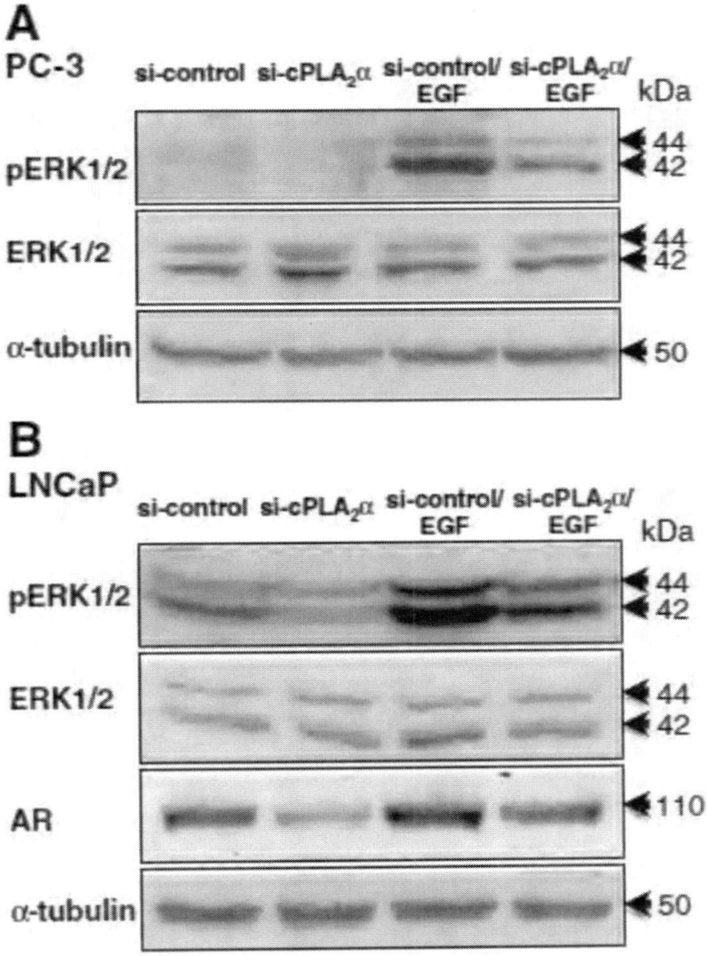

Figure 5. Genetic silencing of $cPLA_2\alpha$ diminishes pERK 1/2 and AR levels in PTEN-null/mutated prostate cancer cells [72]. (A) PC-3 and (B) LNCaP cells transfected with $cPLA_2\alpha$ siRNA for 48 h/37 °C show reduced pERK 1/2 levels relative to the scrambled control under EGF-stimulating (20 ng/mL/30 min) conditions. The level of the AR was reduced under basal and EGF-stimulating conditions after incubation of LNCaP cells with si-$cPLA_2\alpha$ relative to the scrambled control.

We recently determined the effect of decreasing cPLA$_2$α action on both ERK and AKT signalling concomitantly in prostate cancer cells. Genetic silencing or pharmacological inhibition of cPLA$_2$α reduced pERK under basal and EGF-stimulated conditions (Figure 5). Addition of AA significantly increased pERK, and was able to abolish the down-regulation of pERK brought by cPLA$_2$α inhibition (Figure 4). In addition, blocking 5-LOX with MK886 reduced the AA stimulatory effect on pERK, suggesting the role of 5-HETE in mediating AA effect on ERK (Figure 4). Interestingly, while ERK signalling was inhibited by either cPLA$_2$α siRNA or pharmacological agent, no compensational increase in AKT was noted. These findings reveal the importance of cPLA$_2$α, arachidonic acid and eicosanoids in sustaining ERK signalling in prostate cancer cells. The concomitant decrease in both pERK and pAKT levels following inhibition of cPLA$_2$α underscores the importance of cPLA$_2$α in promoting both signalling pathways.

6. Arachidonic Acid and Androgen Receptor Pathway in Prostate Cancer

6.1. The Androgen Receptor (AR) Pathway Overview

The androgen receptor (AR) mediates a wide range of developmental and physiological responses important in male sexual differentiation, maintenance of spermatogenesis and male hormones [92]. The pathway initiates when steroidal androgens, testosterone or its metabolite 5-alpha-dihydrotestosterone (DHT) binds to AR [93, 94], a nuclear receptor [95] normally sequestered into the cytoplasm [96]. Upon binding to androgens, AR translocates from the cytoplasm to the nucleus and regulates expression of androgen-targeted genes [95].

6.2. The Androgen Receptor in Prostate Cancer

AR signalling is able to increase expression of genes for cell cycle and proliferation and is required for PC growth and survival [35, 97-100]. Surgical or medical castration induces apoptosis in the majority of PC cells, which translates clinically to improvement in cancer-related symptoms and lowering of AR-regulated genes, including PSA [35]. PC is initially androgen dependent or responsive to castration, but progresses into castration resistant or hormone refractory, through a mix of molecular and cellular changes resulting tumour growth despite low serum testosterone levels. Treatment options then become limited to suboptimal chemotherapeutic measures, allowing limited lifespan of 1–2 years [35, 101]. Hence, how to block AR signalling pathway other than removal of androgen has become a pressing issue.

6.2.1. Increase in Androgen Receptor Activation and Expression

Up to 50% of prostate tumours have *AR* mutations [102-104]. Mutations in *AR* coding sequence may lead to increased AR activity. These mutations generally transform AR to become 'super receptors', which respond to a wider variety of agonists at lower concentrations. For example, the point-mutation $Thr^{877}\rightarrow Ala$ in the hormone-binding domain allows activation by progesterone, oestradiol, cyproterone acetate and the anti-androgens, anandron, and hydroxyflutamide [105]. Some mutations in the ligand-binding pocket decrease the ligand binding specificity [106, 107], while other mutations change the interaction between various components of the AR transcriptional complex in favour of growth [107].

In addition to mutation in AR gene, the level of AR is an important factor in castration resistance. Increase in AR expression allows utilisation of low levels of androgen, following hormone ablation therapy [108, 109]. Increased AR levels can be due to gene amplification, which is reported in 25–30% of castration resistant PC cases [110-113].

6.2.2. Inhibition of cPLA$_2\alpha$ and Eicosanoid Synthesis Reduces Androgen Receptor Levels

AA plays an important role in both hormone sensitive and castration resistant PC. ROS produced from AA metabolism is able to inhibit PTEN, which is a negative regulator of AR [114]. Dietary AA reduces the time required for conversion of hormone sensitive PC to castration resistant PC after androgen ablation therapy in rodent models [115]. Relapse to castration resistant growth (as measured by renewed increases in tumour volume and serum PSA after androgen ablation) positively correlated with tumour AA content [115]. Thus, AA is critical for the AR pathway throughout PC progression.

We reported recently that there is a clear and strong link between cPLA$_2\alpha$ action and AR levels in PTEN-null prostate cancer cells. Inhibition of cPLA$_2\alpha$ by genetic silencing or pharmacological agent significantly reduced AR expression (Figures 2 and 3). This reduction in AR level was significant following blocking cPLA$_2$α for 24 h. Importantly, the suppressive effect maintained when the cancer cells were stimulated with EGF. cPLA$_2$α inhibition reduced both nuclear and cytosolic AR as well as the secretion of PSA (Figure 6). The effect of cPLA$_2$α inhibition on AR is likely through AA and 5-HETE, because addition of AA significantly increased AR levels in LNCaP cells after 4 h and this was abrogated following incubation with MK886 (Figure 4). Hence, we provide evidence that AR is positively regulated by cPLA$_2$α *via* AA and 5-HETE.

7. cPLA$_2$α Action and Cross-Talk between Oncogenic Pathways

Based on aforementioned studies, AA and its metabolites enhance signalling in AKT, ERK and AR pathways in PC. Blocking AA release by genetic silencing or pharmacological blockade of cPLA$_2$α markedly reduced the signalling in the three oncogenic pathways. It is interesting to note that, although ubiquitously expressed, cPLA$_2$α knockout models exhibited a relatively normal phenotype, unless an inflammation-related pathological process is introduced to challenge cPLA$_2$α knockout animals [20]. These results suggest that cPLA$_2$α is a promising potential therapeutic target and the side effect of blocking cPLA$_2$α in normal tissue may be acceptable.

Conclusion

Cancer is a multi-step process driven by inherited and/or somatic genetic and epigenetic alterations. Thus, to understand cancer aetiology, progression and patient prognosis, it is necessary to identify these alterations and find their links with cancer initiation and progression. Alterations in AA release by increase in cPLA$_2$α activation or AA metabolism is one of the important aspects of PC progression.

AA and eicosanoid regulate signalling in AKT, ERK and AR pathways in PC through multiple avenues: (1) AA and eicosanoids may transactivate EGFR through GPCR to activate PI3K/AKT and ERK [62, 65]; (2) inactivation of PTEN through ROS produced from eicosanoid synthesis leads to increased AKT signalling and removal of AR inhibition [69, 114]; and (3)

AA induces *de novo* androgen synthesis, which is required for AR signalling and confers castration resistance [40].

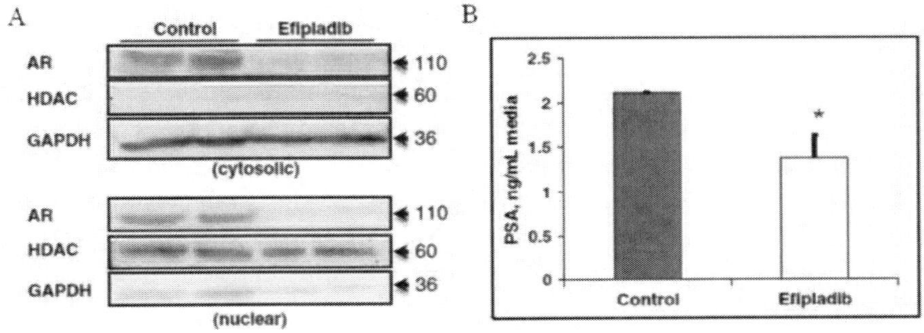

Figure 6. Effects of inhibition of $cPLA_2\alpha$ with Efipladib on AKT, ERK 1/2 and AR levels. LNCaP cells were incubated with Efipladib (20 μM/72 h/37 °C) relative to the vehicle control (0.04% DMSO). (C) The nuclear and cytoplasmic AR levels in LNCaP cell. (D) The PSA levels released into the overlying media of LNCaP cell cultures. Results in (d) are mean ± SD of n = 3 independent experiments. * P< 0.05 versus vehicle-treated control.

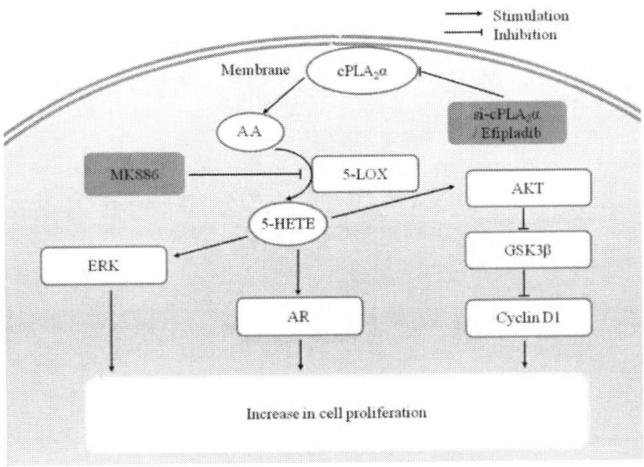

Figure 7. Schematic diagram of $cPLA_2\alpha$ affecting AKT, ERK 1/2 and AR levels. The studies herein demonstrate using genetic approaches that $cPLA_2\alpha$ positively regulates pAKT, pERK 1/2 and AR levels *via* AA and its metabolite 5-HETE derived from the activity of 5-LOX (as confirmed with the 5-LOX pharmacological inhibitor, MK886) in PTEN null/mutated prostate cancer cells. Pharmacological inhibition of $cPLA_2\alpha$ causes reduction in proliferation of PTEN null/mutated prostate cancer cells.

Collectively, $cPLA_2\alpha$/AA play an important role in sustaining AKT, ERK and AR signalling under both basal and EGF-stimulating conditions in PTEN-null/mutated prostate cancer cells (Figure 7). The mechanism of $cPLA_2\alpha$ action on AKT, ERK and AR is likely *via* AA and its metabolite, 5-HETE. Hence, restriction of AA supply to PTEN-null prostate cancer cells *via* inhibition of $cPLA_2\alpha$ could have a therapeutic value by inhibiting constitutively activated AKT signalling. Our study supports the case for restricting dietary AA intake in patients with prostate cancer and for utilising agents blocking $cPLA_2\alpha$ as a novel therapy for this common and deadly disease.

REFERENCES

[1] Lands, B. Consequences of essential fatty acids. *Nutrients*, 2012, 4(9), 1338-57. Epub 2012/11/01.

[2] Jenkins, CM; Cedars, A; Gross, RW. Eicosanoid signalling pathways in the heart. *Cardiovasc Res.*, 2009, 82(2), 240-9. Epub 2008/12/17.

[3] Balsinde, J; Bianco, ID; Ackermann, EJ; Conde-Frieboes, K; Dennis, EA. Inhibition of calcium-independent phospholipase A2 prevents arachidonic acid incorporation and phospholipid remodeling in P388D1 macrophages. *Proc Natl Acad Sci U S A*, 1995, 92(18), 8527-31. Epub 1995/08/29.

[4] Ghosh, M; Tucker, DE; Burchett, SA; Leslie, CC. Properties of the Group IV phospholipase A2 family. *Progress in lipid research*, 2006, 45(6), 487-510. Epub 2006/07/04.

[5] Dessen, A; Tang, J; Schmidt, H; Stahl, M; Clark, JD; Seehra, J; et al., Crystal structure of human cytosolic phospholipase A2 reveals a novel topology and catalytic mechanism. *Cell*, 1999, 97(3), 349-60. Epub 1999/05/13.

[6] Pratico, D. Prostanoid and isoprostanoid pathways in atherogenesis. *Atherosclerosis*, 2008, 201(1), 8-16. Epub 2008/06/03.

[7] Calder, PC. Session 3: Joint Nutrition Society and Irish Nutrition and Dietetic Institute Symposium on 'Nutrition and autoimmune disease' PUFA, inflammatory processes and rheumatoid arthritis. *Proc Nutr Soc*, 2008, 67(4), 409-18. Epub 2008/10/14.

[8] Cunnane, SC. Problems with essential fatty acids: time for a new paradigm? *Progress in lipid research*, 2003, 42(6), 544-68. Epub 2003/10/16.

[9] Patel, MI; Kurek, C; Dong, Q. The arachidonic acid pathway and its role in prostate cancer development and progression. *J Urol*, 2008, 179(5), 1668-75.

[10] Mizenina, O; Musatkina, E; Yanushevich, Y; Rodina, A; Krasilnikov, M; de Gunzburg, J; et al., A novel group IIA phospholipase A2 interacts with v-Src oncoprotein from RSV-transformed hamster cells. *J Biol Chem*, 2001, 276(36), 34006-12. Epub 2001/06/28.

[11] Rouault, M; Bollinger, JG; Lazdunski, M; Gelb, MH; Lambeau, G. Novel mammalian group XII secreted phospholipase A2 lacking enzymatic activity. *Biochemistry*, 2003, 42(39), 11494-503. Epub 2003/10/01.

[12] Six, DA; Dennis, EA. The expanding superfamily of phospholipase A(2) enzymes: classification and characterization. *Biochim Biophys Acta*, 2000, 1488(1-2), 1-19. Epub 2000/11/18.

[13] Scott, KF; Graham, GG; Bryant, KJ. Secreted phospholipase A2 enzymes as therapeutic targets. *Expert opinion on therapeutic targets*, 2003, 7(3), 427-40. Epub 2003/06/05.

[14] Bonventre, JV; Huang, Z; Taheri, MR; O'Leary, E; Li, E; Moskowitz, MA; et al., Reduced fertility and postischaemic brain injury in mice deficient in cytosolic phospholipase A2. *Nature*, 1997, 390(6660), 622-5.

[15] Sapirstein, A; Bonventre, JV. Specific physiological roles of cytosolic phospholipase A(2) as defined by gene knockouts. *Biochim Biophys Acta*, 2000, 1488(1-2), 139-48. Epub 2000/11/18.

[16] Burke, JE; Dennis, EA. Phospholipase A2 structure/function, mechanism, and signaling. *J Lipid Res*, 2009, 50 Suppl(42), S237-42.

[17] Evans, JH; Leslie, CC. The cytosolic phospholipase A2 catalytic domain modulates association and residence time at Golgi membranes. *J Biol Chem*, 2004, 279(7), 6005-16.

[18] Clark, JD; Schievella, AR; Nalefski, EA; Lin, LL. Cytosolic phospholipase A2. *Journal of lipid mediators and cell signalling*, 1995, 12(2-3), 83-117. Epub 1995/10/01.

[19] Hsu, YH; Burke, JE; Stephens, DL; Deems, RA; Li, S; Asmus, KM; et al., Calcium binding rigidifies the C2 domain and the intradomain interaction of GIVA phospholipase A2 as revealed by hydrogen/deuterium exchange mass spectrometry. *J Biol Chem*, 2008, 283(15), 9820-7. Epub 2008/01/24.

[20] Niknami, M; Patel, M; Witting, PK; Dong, Q. Molecules in focus: cytosolic phospholipase A2-alpha. *Int J Biochem Cell Biol*, 2009, 41(5), 994-7.

[21] Ramarao, MK; Shen, MW; Murphy, EA; Duan, W; Zhao, Y; McKew, J; et al., Thermodynamic characterization of cytosolic phospholipase A2 alpha inhibitors. *Anal Biochem*, 2008, 383(2), 217-25. Epub 2008/09/20.

[22] Huang, Z; Payette, P; Abdullah, K; Cromlish, WA; Kennedy, BP. Functional identification of the active-site nucleophile of the human 85-kDa cytosolic phospholipase A2. *Biochemistry*, 1996, 35(12), 3712-21.

[23] Pavicevic, Z; Leslie, CC; Malik, KU. cPLA2 phosphorylation at serine-515 and serine-505 is required for arachidonic acid release in vascular smooth muscle cells. *J Lipid Res*, 2008, 49(4), 724-37.

[24] Shimizu, M; Nakamura, H; Hirabayashi, T; Suganami, A; Tamura, Y; Murayama, T. Ser515 phosphorylation-independent regulation of cytosolic phospholipase A2alpha (cPLA2alpha) by calmodulin-dependent protein kinase: possible interaction with catalytic domain A of cPLA2alpha. *Cell Signal*, 2008, 20(5), 815-24.

[25] Borsch-Haubold, AG; Ghomashchi, F; Pasquet, S; Goedert, M; Cohen, P; Gelb, MH; et al., Phosphorylation of cytosolic phospholipase A2 in platelets is mediated by multiple stress-activated protein kinase pathways. *Eur J Biochem*, 1999, 265(1), 195-203.

[26] Tian, W; Wijewickrama, GT; Kim, JH; Das, S; Tun, MP; Gokhale, N; et al., Mechanism of regulation of group IVA phospholipase A2 activity by Ser727 phosphorylation. *J Biol Chem*, 2008, 283(7), 3960-71.

[27] Hefner, Y; Borsch-Haubold, AG; Murakami, M; Wilde, JI; Pasquet, S; Schieltz, D; et al., Serine 727 phosphorylation and activation of cytosolic phospholipase A2 by MNK1-related protein kinases. *J Biol Chem*, 2000, 275(48), 37542-51. Epub 2000/09/09.

[28] Murakami, M; Kudo, I. Phospholipase A2. *J Biochem*, 2002, 131(3), 285-92. Epub 2002/03/02.

[29] Wilkins, WP 3rd; Barbour, SE. Group VI phospholipases A2: homeostatic phospholipases with significant potential as targets for novel therapeutics. *Curr Drug Targets*, 2008, 9(8), 683-97. Epub 2008/08/12.

[30] Kitsiouli, E; Nakos, G; Lekka, ME. Phospholipase A2 subclasses in acute respiratory distress syndrome. *Biochim Biophys Acta*, 2009, 1792(10), 941-53.

[31] Haas, GP; Sakr, WA. Epidemiology of prostate cancer. *CA Cancer J Clin*, 1997, 47(5), 273-87. Epub 1997/10/07.

[32] PCFA. What is Prostate Cancer Available from: http: //wwwprostateorgau/articleLive/pages/What-is-Prostate-Cancerhtml. 2012 [citied 2012 11 June].

[33] Garcia, M; Jemal, A; Ward, EM; Center, MM; Hao, Y; Siegel, RL; et al., *Global Cancer Facts & Figures 2007*. Atlanta, GA: American Cancer Society, 2007.

[34] Adolfsson, J. Watchful waiting and active surveillance: the current position. *BJU Int*, 2008, 102(1), 10-4. Epub 2008/04/22.

[35] Taplin, ME; Balk, SP. Androgen receptor: a key molecule in the progression of prostate cancer to hormone independence. *J Cell Biochem*, 2004, 91(3), 483-90. Epub 2004/02/03.

[36] McEntee, MF; Ziegler, C; Reel, D; Tomer, K; Shoieb, A; Ray, M; et al., Dietary n-3 polyunsaturated fatty acids enhance hormone ablation therapy in androgen-dependent prostate cancer. *American Journal of Pathology*, 2008, 173(1), 229-41.

[37] Kelavkar, UP; Hutzley, J; Dhir, R; Kim, P; Allen, KG; McHugh, K; et al., Prostate tumor growth and recurrence can be modulated by the omega-6: omega-3 ratio in diet: athymic mouse xenograft model simulating radical prostatectomy. Neoplasia (New York), 2006, 8(2), 112-24.

[38] Berquin, IM; Min, Y; Wu, R; Wu, J; Perry, D; Cline, JM; et al., Modulation of prostate cancer genetic risk by omega-3 and omega-6 fatty acids. *Journal of Clinical Investigation*, 2007, 117(7), 1866-75.

[39] Ghosh, J; Myers, CE. Arachidonic acid stimulates prostate cancer cell growth: critical role of 5-lipoxygenase. *Biochemical & Biophysical Research Communications*, 1997, 235(2), 418-23.

[40] Locke, JA; Guns, ES; Lehman, ML; Ettinger, S; Zoubeidi, A; Lubik, A; et al., Arachidonic acid activation of intratumoral steroid synthesis during prostate cancer progression to castration resistance. *Prostate*, 2010, 70(3), 239-51. Epub 2009/10/01.

[41] Chaudry, AA; Wahle, KW; McClinton, S; Moffat, LE. Arachidonic acid metabolism in benign and malignant prostatic tissue in vitro: effects of fatty acids and cyclooxygenase inhibitors. *International Journal of Cancer*, 1994, 57(2), 176-80.

[42] Pruthi, RS; Derksen, JE; Moore, D; Carson, CC; Grigson, G; Watkins, C; et al., Phase II trial of celecoxib in prostate-specific antigen recurrent prostate cancer after definitive radiation therapy or radical prostatectomy. *Clinical Cancer Research*, 2006, 12((7 Pt 1)), 2172-7.

[43] Smith, MR; Manola, J; Kaufman, DS; Oh, WK; Bubley, GJ; Kantoff, PW. Celecoxib versus placebo for men with prostate cancer and a rising serum prostate-specific antigen after radical prostatectomy and/or radiation therapy.[see comment]. *Journal of Clinical Oncology*, 2006, 24(18), 2723-8.

[44] Sved, P; Scott, KF; McLeod, D; King, N; Singh, J; Tsatralis, T; et al., Oncogenic action of secreted phospholipase A2 in prostate cancer. *Cancer Research*, 2004, 64(19), 6934-40.

[45] Patel, MI; Singh, J; Niknami, M; Kurek, C; Yao, M; Lu, S; et al., Cytosolic phospholipase A2-alpha: a potential therapeutic target for prostate cancer. *Clinical Cancer Research*, 2008, 14(24), 8070-9.

[46] Chetcuti, A; Margan, SH; Russell, P; Mann, S; Millar, DS; Clark, SJ; et al., Loss of Annexin II heavy and light chains in prostate cancer and its precursors. *Cancer Research*, 2001, 61(17), 6331-4.

[47] Dong, Q; Patel, M; Scott, KF; Graham, GG; Russell, PJ; Sved, P. Oncogenic action of phospholipase A2 in prostate cancer. *Cancer Letters*, 2006, 240(1), 9-16.

[48] Chalhoub, N; Baker, SJ. PTEN and the PI3-kinase pathway in cancer. *Annu Rev Pathol*, 2009, 4, 127-50.

[49] Engelman, JA; Luo, J; Cantley, LC. The evolution of phosphatidylinositol 3-kinases as regulators of growth and metabolism. *Nat Rev Genet*, 2006, 7(8), 606-19.

[50] Osaki, M; Oshimura, M; Ito, H. PI3K-Akt pathway: its functions and alterations in human cancer. *Apoptosis: an international journal on programmed cell death*, 2004, 9(6), 667-76. Epub 2004/10/27.

[51] Yuan, TL; Cantley, LC. PI3K pathway alterations in cancer: variations on a theme. *Oncogene*, 2008, 27(41), 5497-510. Epub 2008/09/17.

[52] Cuevas, BD; Lu, Y; Mao, M; Zhang, J; LaPushin, R; Siminovitch, K; et al., Tyrosine phosphorylation of p85 relieves its inhibitory activity on phosphatidylinositol 3-kinase. *J Biol Chem*, 2001, 276(29), 27455-61. Epub 2001/05/05.

[53] Dhand, R; Hiles, I; Panayotou, G; Roche, S; Fry, MJ; Gout, I; et al., PI 3-kinase is a dual specificity enzyme: autoregulation by an intrinsic protein-serine kinase activity. *Embo J*, 1994, 13(3), 522-33. Epub 1994/02/01.

[54] Carpenter, CL; Auger, KR; Duckworth, BC; Hou, WM; Schaffhausen, B; Cantley, LC. A tightly associated serine/threonine protein kinase regulates phosphoinositide 3-kinase activity. *Mol Cell Biol*, 1993, 13(3), 1657-65. Epub 1993/03/01.

[55] Assinder, SJ; Dong, Q; Kovacevic, Z; Richardson, DR. The TGF-beta, PI3K/Akt and PTEN pathways: established and proposed biochemical integration in prostate cancer. *Biochem J*, 2009, 417(2), 411-21.

[56] Vanhaesebroeck, B; Leevers, SJ; Ahmadi, K; Timms, J; Katso, R; Driscoll, PC; et al., Synthesis and function of 3-phosphorylated inositol lipids. *Annual review of biochemistry*, 2001, 70, 535-602. Epub 2001/06/08.

[57] Bellacosa, A; Kumar, CC; Di Cristofano, A; Testa, JR. Activation of AKT kinases in cancer: implications for therapeutic targeting. *Adv Cancer Res*, 2005, 94, 29-86.

[58] O'Reilly, KE; Rojo, F; She, QB; Solit, D; Mills, GB; Smith, D; et al., mTOR inhibition induces upstream receptor tyrosine kinase signaling and activates Akt. *Cancer Res*, 2006, 66(3), 1500-8.

[59] Um, SH; Frigerio, F; Watanabe, M; Picard, F; Joaquin, M; Sticker, M; et al., Absence of S6K1 protects against age- and diet-induced obesity while enhancing insulin sensitivity. *Nature*, 2004, 431(7005), 200-5. Epub 2004/08/13.

[60] Zeng, ZZ; Yellaturu, CR; Neeli, I; Rao, GN. 5(S)-hydroxyeicosatetraenoic acid stimulates DNA synthesis in human microvascular endothelial cells via activation of Jak/STAT and phosphatidylinositol 3-kinase/Akt signaling, leading to induction of expression of basic fibroblast growth factor 2. *J Biol Chem*, 2002, 277(43), 41213-9. Epub 2002/08/24.

[61] Guo, AM; Liu, X; Al-Wahab, Z; Maddippati, KR; Ali-Fehmi, R; Scicli, AG; et al., Role of 12-lipoxygenase in regulation of ovarian cancer cell proliferation and survival. *Cancer Chemother Pharmacol*, 2011, 68(5), 1273-83. Epub 2011/03/29.

[62] Vo, BT; Morton, D Jr.; Komaragiri, S; Millena, AC; Leath, C; Khan, SA. TGF-beta effects on prostate cancer cell migration and invasion are mediated by PGE2 through activation of PI3K/AKT/mTOR pathway. *Endocrinology*, 2013, 154(5), 1768-79. Epub 2013/03/22.

[63] Tawadros, T; Brown, MD; Hart, CA; Clarke, NW. Ligand-independent activation of EphA2 by arachidonic acid induces metastasis-like behaviour in prostate cancer cells. *British journal of cancer*, 2012, 107(10), 1737-44. Epub 2012/10/06.

[64] Hughes-Fulford, M; Li, CF; Boonyaratanakornkit, J; Sayyah, S. Arachidonic acid activates phosphatidylinositol 3-kinase signaling and induces gene expression in prostate cancer. *Cancer Res*, 2006, 66(3), 1427-33.

[65] Tveteraas, IH; Muller, KM; Aasrum, M; Odegard, J; Dajani, O; Guren, T; et al., Mechanisms involved in PGE2-induced transactivation of the epidermal growth factor receptor in MH1C1 hepatocarcinoma cells. *J Exp Clin Cancer Res*, 2012, 31, 72. Epub 2012/09/13.

[66] Hassan, S; Carraway, RE. Involvement of arachidonic acid metabolism and EGF receptor in neurotensin-induced prostate cancer PC3 cell growth. *Regul Pept*, 2006, 133(1-3), 105-14.

[67] Li, J; Yen, C; Liaw, D; Podsypanina, K; Bose, S; Wang, SI; et al., PTEN, a putative protein tyrosine phosphatase gene mutated in human brain, breast, and prostate cancer. *Science*, 1997, 275(5308), 1943-7. Epub 1997/03/28.

[68] Vlietstra, RJ; van Alewijk, DC; Hermans, KG; van Steenbrugge, GJ; Trapman, J. Frequent inactivation of PTEN in prostate cancer cell lines and xenografts. *Cancer Res*, 1998, 58(13), 2720-3. Epub 1998/07/14.

[69] Covey, TM; Edes, K; Fitzpatrick, FA. Akt activation by arachidonic acid metabolism occurs via oxidation and inactivation of PTEN tumor suppressor. *Oncogene*, 2007, 26(39), 5784-92.

[70] Seo, JH; Ahn, Y; Lee, SR; Yeol Yeo, C; Chung Hur, K. The major target of the endogenously generated reactive oxygen species in response to insulin stimulation is phosphatase and tensin homolog and not phosphoinositide-3 kinase (PI-3 kinase) in the PI-3 kinase/Akt pathway. *Mol Biol Cell*, 2005, 16(1), 348-57. Epub 2004/11/13.

[71] Aggarwal, BB; Shishodia, S; Sandur, SK; Pandey, MK; Sethi, G. Inflammation and cancer: how hot is the link? *Biochem Pharmacol*, 2006, 72(11), 1605-21.

[72] Hua, S; Yao, M; Vignarajan, S; Witting, P; Hejazi, L; Gong, Z; et al., Cytosolic phospholipase A2alpha sustains pAKT, pERK and AR levels in PTEN-null/mutated prostate cancer cells. *Biochim Biophys Acta*, 2013, 1831(6), 1146-57. Epub 2013/03/19.

[73] Pidgeon, GP; Lysaght, J; Krishnamoorthy, S; Reynolds, JV; O'Byrne, K; Nie, D; et al., Lipoxygenase metabolism: roles in tumor progression and survival. *Cancer metastasis reviews*, 2007, 26(3-4), 503-24. Epub 2007/10/19.

[74] Moreno, JJ. New aspects of the role of hydroxyeicosatetraenoic acids in cell growth and cancer development. *Biochem Pharmacol*, 2009, 77(1), 1-10. Epub 2008/09/02.

[75] Subramaniam, S; Unsicker, K. Extracellular signal-regulated kinase as an inducer of non-apoptotic neuronal death. *Neuroscience*, 2006, 138(4), 1055-65. Epub 2006/01/31.

[76] Wada, T; Penninger, JM. Mitogen-activated protein kinases in apoptosis regulation. *Oncogene*, 2004, 23(16), 2838-49. Epub 2004/04/13.

[77] Steelman, LS; Pohnert, SC; Shelton, JG; Franklin, RA; Bertrand, FE; McCubrey, JA. JAK/STAT, Raf/MEK/ERK, PI3K/Akt and BCR-ABL in cell cycle progression and leukemogenesis. *Leukemia*, 2004, 18(2), 189-218. Epub 2004/01/23.

[78] McCubrey, JA; Steelman, LS; Chappell, WH; Abrams, SL; Wong, EW; Chang, F; et al., Roles of the Raf/MEK/ERK pathway in cell growth, malignant transformation and drug resistance. *Biochim Biophys Acta*, 2007, 1773(8), 1263-84. Epub 2006/11/28.

[79] Roberts, PJ; Der, CJ. Targeting the Raf-MEK-ERK mitogen-activated protein kinase cascade for the treatment of cancer. *Oncogene*, 2007, 26(22), 3291-310. Epub 2007/05/15.

[80] Calvo, F; Agudo-Ibanez, L; Crespo, P. The Ras-ERK pathway: understanding site-specific signaling provides hope of new anti-tumor therapies. *Bioessays*, 2010, 32(5), 412-21.

[81] Zarich, N; Oliva, JL; Martinez, N; Jorge, R; Ballester, Ap; Gutierrez-Eisman, S; et al., Grb2 is a negative modulator of the intrinsic Ras-GEF activity of hSos1. *Mol Biol Cell*, 2006, 17(8), 3591-7. Epub 2006/06/09.

[82] Avruch, J; Khokhlatchev, A; Kyriakis, JM; Luo, Z; Tzivion, G; Vavvas, D; et al., Ras activation of the Raf kinase: tyrosine kinase recruitment of the MAP kinase cascade. *Recent Prog Horm Res*, 2001, 56, 127-55. Epub 2001/03/10.

[83] Gioeli, D; Mandell, JW; Petroni, GR; Frierson, HF; Jr. Weber, MJ. Activation of mitogen-activated protein kinase associated with prostate cancer progression. *Cancer Res*, 1999, 59(2), 279-84. Epub 1999/02/02.

[84] Abreu-Martin, MT; Chari, A; Palladino, AA; Craft, NA; Sawyers, CL. Mitogen-activated protein kinase kinase kinase 1 activates androgen receptor-dependent transcription and apoptosis in prostate cancer. *Mol Cell Biol*, 1999, 19(7), 5143-54. Epub 1999/06/22.

[85] Weber, MJ; Gioeli, D. Ras signaling in prostate cancer progression. *J Cell Biochem*, 2004, 91(1), 13-25. Epub 2003/12/23.

[86] Bakin, RE; Gioeli, D; Sikes, RA; Bissonette, EA; Weber, MJ. Constitutive activation of the Ras/mitogen-activated protein kinase signaling pathway promotes androgen hypersensitivity in LNCaP prostate cancer cells. *Cancer Res*, 2003, 63(8), 1981-9. Epub 2003/04/19.

[87] Mukherjee, R; Bartlett, JM; Krishna, NS; Underwood, MA; Edwards, J. Raf-1 expression may influence progression to androgen insensitive prostate cancer. *Prostate*, 2005, 64(1), 101-7. Epub 2005/01/25.

[88] Malik, SN; Brattain, M; Ghosh, PM; Troyer, DA; Prihoda, T; Bedolla, R; et al., Immunohistochemical demonstration of phospho-Akt in high Gleason grade prostate cancer. *Clin Cancer Res*, 2002, 8(4), 1168-71.

[89] Kreisberg, JI; Malik, SN; Prihoda, TJ; Bedolla, RG; Troyer, DA; Kreisberg, S; et al., Phosphorylation of Akt (Ser473) is an excellent predictor of poor clinical outcome in prostate cancer. *Cancer Res*, 2004, 64(15), 5232-6. Epub 2004/08/04.

[90] Kalaany, NY; Sabatini, DM. Tumours with PI3K activation are resistant to dietary restriction. *Nature*, 2009, 458(7239), 725-31. Epub 2009 Mar 11.

[91] Grant, S. Cotargeting survival signaling pathways in cancer. *J Clin Invest*, 2008, 118(9), 3003-6.

[92] Heinlein, CA; Chang, C. The roles of androgen receptors and androgen-binding proteins in nongenomic androgen actions. *Mol Endocrinol*, 2002, 16(10), 2181-7. Epub 2002/09/28.

[93] Roy, AK; Lavrovsky, Y; Song, CS; Chen, S; Jung, MH; Velu, NK; et al., Regulation of androgen action. *Vitam Horm*, 1999, 55, 309-52. Epub 1999/02/09.

[94] Lee, DK; Chang, C. Molecular communication between androgen receptor and general transcription machinery. *J Steroid Biochem Mol Biol*, 2003, 84(1), 41-9. Epub 2003/03/22.

[95] Lu, NZ; Wardell, SE; Burnstein, KL; Defranco, D; Fuller, PJ; Giguere, V; et al., International Union of Pharmacology. LXV. The pharmacology and classification of the nuclear receptor superfamily: glucocorticoid, mineralocorticoid, progesterone, and androgen receptors. *Pharmacol Rev*, 2006, 58(4), 782-97. Epub 2006/11/30.

[96] Friedlander, TW; Ryan, CJ. Targeting the androgen receptor. *Urol Clin North Am*, 2012, 39(4), 453-64. Epub 2012/10/23.

[97] Culig, Z; Klocker, H; Bartsch, G; Steiner, H; Hobisch, A. Androgen receptors in prostate cancer. *J Urol*, 2003, 170(4 Pt 1), 1363-9. Epub 2003/09/23.

[98] Debes, JD; Tindall, DJ. Mechanisms of androgen-refractory prostate cancer. *The New England journal of medicine*, 2004, 351(15), 1488-90. Epub 2004/10/08.

[99] Grossmann, ME; Huang, H; Tindall, DJ. Androgen receptor signaling in androgen-refractory prostate cancer. *Journal of the National Cancer Institute*, 2001, 93(22), 1687-97. Epub 2001/11/22.

[100] Scher, HI; Sawyers, CL. Biology of progressive, castration-resistant prostate cancer: directed therapies targeting the androgen-receptor signaling axis. *J Clin Oncol*, 2005, 23(32), 8253-61. Epub 2005/11/10.

[101] Savarese, DM; Halabi, S; Hars, V; Akerley, WL; Taplin, ME; Godley, PA; et al., Phase II study of docetaxel, estramustine, and low-dose hydrocortisone in men with hormone-refractory prostate cancer: a final report of CALGB 9780. Cancer and Leukemia Group B. *J Clin Oncol*, 2001, 19(9), 2509-16. Epub 2001/05/02.

[102] Taplin, ME; Bubley, GJ; Shuster, TD; Frantz, ME; Spooner, AE; Ogata, GK; et al., Mutation of the androgen-receptor gene in metastatic androgen-independent prostate cancer. *The New England journal of medicine*, 1995, 332(21), 1393-8. Epub 1995/05/25.

[103] Suzuki, H; Sato, N; Watabe, Y; Masai, M; Seino, S; Shimazaki, J. Androgen receptor gene mutations in human prostate cancer. *J Steroid Biochem Mol Biol*, 1993, 46(6), 759-65. Epub 1993/12/01.

[104] Marcelli, M; Ittmann, M; Mariani, S; Sutherland, R; Nigam, R; Murthy, L; et al., Androgen receptor mutations in prostate cancer. *Cancer Res*, 2000, 60(4), 944-9. Epub 2000/03/08.

[105] Veldscholte, J; Ris-Stalpers, C; Kuiper, GG; Jenster, G; Berrevoets, C; Claassen, E; et al., A mutation in the ligand binding domain of the androgen receptor of human LNCaP cells affects steroid binding characteristics and response to anti-androgens. *Biochem Biophys Res Commun*, 1990, 173(2), 534-40. Epub 1990/12/14.

[106] Buchanan, G; Greenberg, NM; Scher, HI; Harris, JM; Marshall, VR; Tilley, WD. Collocation of androgen receptor gene mutations in prostate cancer. *Clin Cancer Res*, 2001, 7(5), 1273-81. Epub 2001/05/15.

[107] Gelmann, EP. Molecular biology of the androgen receptor. *J Clin Oncol*, 2002, 20(13), 3001-15. Epub 2002/06/29.

[108] Chen, CD; Welsbie, DS; Tran, C; Baek, SH; Chen, R; Vessella, R; et al., Molecular determinants of resistance to antiandrogen therapy. *Nature medicine*, 2004, 10(1), 33-9. Epub 2004/01/02.

[109] Yuan, X; Balk, SP. Mechanisms mediating androgen receptor reactivation after castration. *Urol Oncol*, 2009, 27(1), 36-41. Epub 2008/12/30.

[110] Visakorpi, T; Hyytinen, E; Koivisto, P; Tanner, M; Keinanen, R; Palmberg, C; et al., In vivo amplification of the androgen receptor gene

and progression of human prostate cancer. *Nature genetics*, 1995, 9(4), 401-6. Epub 1995/04/01.

[111] Koivisto, P; Kononen, J; Palmberg, C; Tammela, T; Hyytinen, E; Isola, J; et al., Androgen receptor gene amplification: a possible molecular mechanism for androgen deprivation therapy failure in prostate cancer. *Cancer Res*, 1997, 57(2), 314-9. Epub 1997/01/15.

[112] Miyoshi, Y; Uemura, H; Fujinami, K; Mikata, K; Harada, M; Kitamura, H; et al., Fluorescence in situ hybridization evaluation of c-myc and androgen receptor gene amplification and chromosomal anomalies in prostate cancer in Japanese patients. *Prostate*, 2000, 43(3), 225-32. Epub 2000/05/08.

[113] Linja, MJ; Savinainen, KJ; Saramaki, OR; Tammela, TL; Vessella, RL; Visakorpi, T. Amplification and overexpression of androgen receptor gene in hormone-refractory prostate cancer. *Cancer Res*, 2001, 61(9), 3550-5. Epub 2001/04/28.

[114] Lin, HK; Hu, YC; Lee, DK; Chang, C. Regulation of androgen receptor signaling by PTEN (phosphatase and tensin homolog deleted on chromosome 10) tumor suppressor through distinct mechanisms in prostate cancer cells. *Mol Endocrinol*, 2004, 18(10), 2409-23.

[115] McEntee, MF; Ziegler, C; Reel, D; Tomer, K; Shoieb, A; Ray, M; et al., Dietary n-3 polyunsaturated fatty acids enhance hormone ablation therapy in androgen-dependent prostate cancer. *Am J Pathol*, 2008, 173(1), 229-41. Epub 2008 Jun 13.

In: Arachidonic Acid
Editor: Jason M. O'Keefe
ISBN: 978-1-63117-619-7

Chapter 6

EFFECTS OF ADIPOKINES ON PROSTAGLANDIN E2 PRODUCTION BY RHEUMATOID SYNOVIAL FIBROBLASTS

Natsuko Kusunoki[1], Fumiaki Kojima[2] and Shinichi Kawai[1*]

[1]Division of Rheumatology, Department of Internal Medicine, Toho University School of Medicine, Japan
[2]Department of Pharmacology, Asahikawa Medical University, Japan

ABSTRACT

Eicosanoids, including prostaglandins (PGs), leukotrienes (LTs), and lipoxins (LXs), regulate a wide variety of physiological responses and pathological processes. Among them, PGE_2 is one of the key molecules, especially in relation to inflammation and immunity. Rheumatoid arthritis (RA) is characterized by extensive inflammation and proliferation of the synovium in various joints of the body. Since proinflammatory cytokines, such as tumor necrosis factor-α (TNFα) and interleukin (IL)-1β, induce PGE_2 production by synovial fibroblasts, joint inflammation in RA is at least partly mediated by overproduction of PGE_2. We previously reported

* Corresponding author: Shinichi Kawai, MD, PhD, Professor of Internal Medicine, Division of Rheumatology, School of Medicine, Faculty of Medicine, Toho University, 6-11-1 Omori-Nishi, Ota-ku, Tokyo 143-8541, Japan, Tel. +81-3-3762-4151 (ext. 6591)Fax. +81-3-5753-8513. E-mail.skawai@med.toho-u.ac.jp.

that PGE_2 overproduction led to a vicious cycle of inflammation because PGE_2 itself promotes the increased expression of microsomal prostaglandin E synthase-1 (mPGES-1), a downstream enzyme of cyclooxygenase-2 (COX-2).

Adipose tissue has long been considered to be merely a structural component of many organs and a site for energy storage. However, recent studies have demonstrated that the major cellular component of adipose tissue, the adipocyte, has the ability to synthesize and release various physiologically active molecules, including diponectin, leptin, and resistin, as well as well-known proinflammatory cytokines like TNFα and IL-6. These molecules are called adipokines. We recently found that adiponectin stimulates PGE_2 production by increasing the expression of COX-2 and mPGES-1 in rheumatoid synovial fibroblasts.

In this review, we summarize the significance of adipokines with regard to PGE_2 and rheumatoid inflammation. We also suggest that adiponectin might have a role as a proinflammatory cytokine in rheumatoid arthritis.

PROINFLAMMATORY CYTOKINES IN RHEUMATOID ARTHRITIS

Rheumatoid arthritis (RA) is characterized by extensive inflammation and proliferation of the synovium in multiple joints. Since proinflammatory cytokines, such as tumor necrosis factor-α (TNFα), interleukin (IL)-1β, and IL-6, play a central role in the pathophysiologic mechanisms underlying RA, novel strategies to neutralize these cytokines by employing monoclonal antibodies or soluble receptors have recently been developed as new treatments for RA [1].

Although the pathogenesis of RA is still unknown, the intricate network of proinflammatory cytokines involved in this disease has been studied extensively. Various secreted cytokines stimulate other cells to produce various bioactive molecules, including metabolites of arachidonic acid (AA), resultingin the inflammatory and destructive processes of RA.

Cytokines and Adipokines

Adipose tissue was long considered to be just a structural component of many organs and a site for energy storage. However, some recent studies have shown that the adipocyte (the major cellular component of adipose tissue) can synthesize and release physiologically active molecules or cytokines, including adiponectin, leptin, and resistin, as well as conventional cytokines such as IL-6 and TNFα [2]. These molecules are collectively called adipokines.

Although blockade of TNFα, IL-1β, or IL-6 is clinically beneficial, such strategies are not curative and the effect is only partial, with failure to respond being common [1]. Therefore, it seems possible that other proinflammatory cytokines may contribute to the pathophysiology of RA.

It has been suggested that adiponectin may play a central role in the regulation of insulin resistance [3]. However, adiponectin is also involved in many aspects of inflammation and immunity [4, 5]. We found that the serum levels of leptin and adiponectin were elevated in patients with RA [6]. Other studies have also shown that serum levels of resistin [7-9], leptin [9, 11-13], and adiponectin [9, 10, 13-15] are higher in RA patients than in healthy controls. These results suggest that some adipokines have a rolein the pathophysiological process of RA as doproinflammatory cytokines.

AA Cascade in RA

Phospholipase A2

Figure1 shows the AA cascade. Biosynthesis of eicosanoids, including prostaglandins (PGs), thromboxanes (TXs), leukotrienes (LTs) and lipoxins (LXs), depends on the availability of free AA [16]. When tissues are exposed to physiological and pathological stimuli, such as growth factors, hormones, or cytokines, AA is produced from membrane phospholipids by the action of phospholipase A2s (PLA2s). PLA2s show structural diversity. Among them, cytosolic phospholipase A2α (cPLA2α) has been well characterized with respect to its protein structure and properties, and it is thought to play an essential role in eicosanoid production [17, 18]. cPLA2α undergoes translocation from the cytosol to the perinuclear membrane when stimulated by intracellular Ca^{2+} at submicromolar levels [19], after which it hydrolyzes AA-containing phospholipids and supplies AA to downstream enzymes such as cyclooxygenase (COX) and lipoxygenase (LOX). These enzymes produce various eicosanoids.

COX

AA can be metabolized by three main enzymatic pathways, which are p-450 epoxygenase, COXs, and LOXs (Figure1).

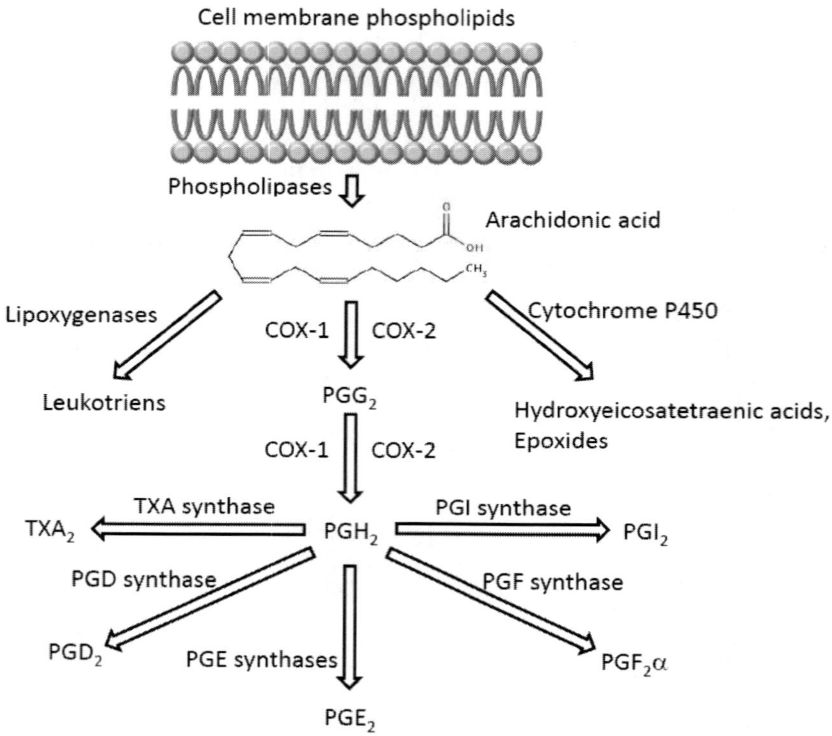

Figure1. Arachidonic acid cascade. COX, cyclooxygenase; PG, prostaglandin; TX, thromboxane.

COX has 2 isozymes, COX-1 and COX-2, which differ with respect to their basal level of expression, tissue localization, and induction during inflammation [20-22].

COX-1 is constitutively expressed by various cells and tissues, and has an important role in maintaining homeostasis. In contrast, COX-2 expression is induced in inflammatory cells and tissues by various stimuli, suggesting that it has a key role in the process of inflammation. Nonsteroidal anti-inflammatory drugs (NSAIDs) have both therapeutic and toxic effects that are mediated by their ability to reduce PG biosynthesis through inhibition of COXs [23].

COX-2 mRNA expression was reported to be significantly higher in synovial tissue from RA patients than in tissue from osteoarthritis (OA) patients [24]. It has also been reported that cytokine-activated cells, such as synovial cells, chondrocytes and macrophages/monocytes, are the primary source of PGs in arthritic joints. Furthermore, PG production at sites of inflammation coincides with the upregulation of COX-2 expression in activated articular cells [25].

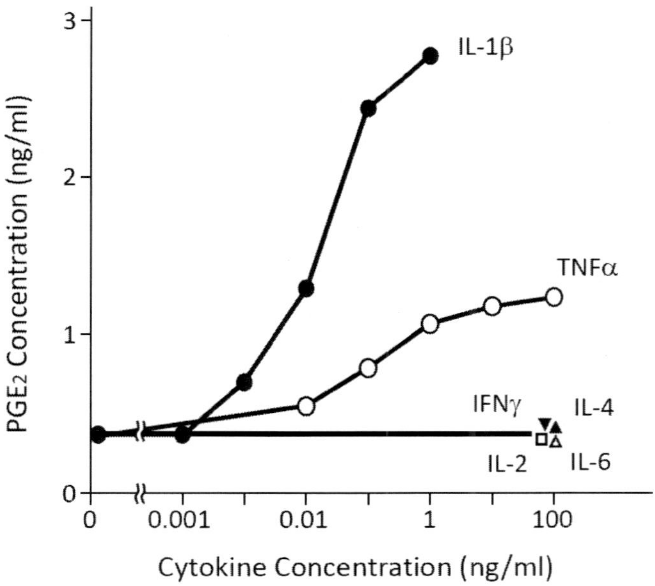

Figure2. Effects of various cytokines on COX activity in human synovial cells. Synovial cells were treated with cytokines at 37°C for 24 h. After washing the cells, 3 μM arachidonic acid was added for 30 min and then the prostaglandin E_2 (PGE_2) concentration of the medium was measured. Each point represents the mean value of three samples. IL, interleukin; TNFα, tumor necrosis factor-a; IFNγ, interferon-γ. Reproduced from Kawai et al., [28], with kind permission from European Journal of Pharmacology.

PGE_2 in RA

AA is transformed into PGG_2 by COX, and then is further catalyzed to an unstable intermediate (PGH_2) by the peroxidase activity of COX. PGH_2 is a substrate for the production of PGE_2, PGD_2, $PGF_2\alpha$, PGI_2, and TXA_2 (Figure1). Among them, PGE_2 is a key mediator of immunopathology [26]. High concentrations of PGE_2 have been detected in the synovial fluid of patients with RA [27], and IL-1β and TNFα have been shown to induce PGE_2 production by RA synovial fibroblastsin a dose-dependent manner [28]. In contrast, IL-2, IL-4, IL-6, and interferon-γ (IFNγ) have no influence on PGE_2 production by human synovial cells (Figure 2). Since IL-1β and TNFα are key cytokines associated with the pathophysiology of RA, induction of COX and the resulting over production of PGE_2 induced by these cytokines may play an important role in rheumatoid synovial inflammation. PGE_2 has numerous biological actions (Table 1), some of which contribute to joint destruction in RA.

Table 1. Effects of PGE_2 on joint inflammation and destruction

Outcome	Mechanism	Reference
Pain	Reduction of the temperature threshold for TRPV1 activation	[62]
	Activation of the TTX-R Na^+ channel	[63]
	Induction of hypersensitivity via voltage-gated sodium channel Nav1.9	[64]
	Activation of the purinergic P_2X_3 receptor channel	[65]
	Facilitating spinal release of excitatory neurotransmitters	[66,67]
	Enhancement of AMPAR and NMDAR	[68]
	Blocking the glycinergic receptor α3 subunit	[69]
Edema	Increasing vascular permeability	[70]
	Regulation of vascular tone	[71]
Bone destruction	Upregulation of RANKL expression	[72-76]
	Inhibition of OPG expression	[77,78]
	Induction of chondrocyte death	[79]
	Induction of MMPs	[80-82]
	Induction of ADAMTS1	[83]

ADAMTS1, a disintegrin and metallopeptidase with thrombospondin type 1;AMPAR, 2-amino-3-(5-methyl-3-oxo-1,2-oxasol-4-yl) propanoic acid receptor; MMPs, matrix metalloproteinases; NMDAR, N-methyl-$_D$-aspartic acid receptor; OPG, osteoprotegerin; RANKL, receptor activator of NFκB ligand;TRPV1, transient receptor potential subtype V1; TTX-R, tetrodotoxin-resistant.

PGE Synthase (PGES)

PGES is a key enzyme involved in the biosynthesis of PGE_2 that acts downstream of COX. So far, at least three forms of PGES have been cloned and characterized. These are known as cytosolic PGES (cPGES), microsomal prostaglandin E synthase-1 (mPGES-1), and mPGES-2.

cPGES was originally identified as heat shock protein 90 (Hsp90)-associated protein p23 [29]. cPGES is constitutively expressed in the cytosol under basal conditions in a wide variety of cells and tissues. cPGES is functionally coupled with COX-1 and may contribute to the physiological production of PGE_2 for maintenance of homeostasis. Interestingly, mPGES-1 shows coordinated induction with COX-2 by various inflammatory stimuli in cancer cells [30], macrophages [31, 32], fibroblasts [33-35], osteoblasts [36], and chondrocytes [37]. mPGES-1 is preferentially linked with COX-2 rather than COX-1 [31]. Recent reports have demonstrated that mPGES-1 plays an important role in various pathophysiological events related to a wide variety of conditions and processes, including pain [38-40], fever [41-44], bone resorption [36], vascular angiogenesis [39], reproduction [45, 46], Alzheimer's disease [47], atherosclerosis [48, 49], and tumorigenesis [50].

In the RA synovium, mPGES-1 is localized in synovial lining cells, mononuclear and fibroblast-like cells from the sublining, infiltrating synovial macrophages, and vascular endothelial cells [37]. Moreover, mPGES-1 expression isselectively upregulated in the synoviumwhen RA is active, while its expression insynovial tissue is minimal when the disease is inactive [37, 51]. These findings suggest that induction of mPGES-1 in the synovial tissue of patients with active RA plays a major role in PGE_2-related joint inflammation.

Because mPGES-2 is constitutively expressed in various cells/tissues and is functionally coupled with both COX-1 and COX-2, it could have a role in PGE_2 production related to both homeostasis (with COX-1) and inflammation (with COX-2). However, mPGES-2 immunoreactivity has been detected in synovial lining cells and sublining cells, and it is similarly detected in both active and resting RA tissues [51]. Thus, mPGES-2 is probably associated with basal production of PGE_2 that is unrelated to inflammation.

PGE_2-Induced Upregulation of mPGES-1

Our previous study showed that NSAIDs, such as selective COX-2 inhibitors, decrease the expression of mPGES-1 by IL-1β-stimulated RA synovial fibroblasts [52], with this inhibition being overcome by addition of PGE_2 itself. Various physiological effects of PGE_2 are mediated via the EP receptor, which has four subtypes (EP1, EP2, EP3, and EP4). Among these, the EP2 and EP4 receptors increase cyclic AMP via activation of adenylate cyclase. We detected EP2 and EP4 receptors in the synovial fibroblasts of RA patients. In addition to PGE_2, selective EP2 and EP4 receptor agonists increase mPGES-1 expression as does an activator of adenylate cyclase (forskolin), suggesting that PGE_2 enhances the expression of mPGES-1 in rheumatoid synovial fibroblasts by increasing cAMP through activation of the EP2 and EP4 receptors.

We also observed similar findings in chondrocytes. An outline of the positive feedback mechanism regulating mPGES-1 expression is shown in Figure 3. In conclusion, PGE_2 is a strong enhancer of the induction of mPGES-1 expression by IL-1β. Autoregulation of mPGES-1 expression by PGE_2 may play an important role in the vicious circle of inflammation associated with arthritis.

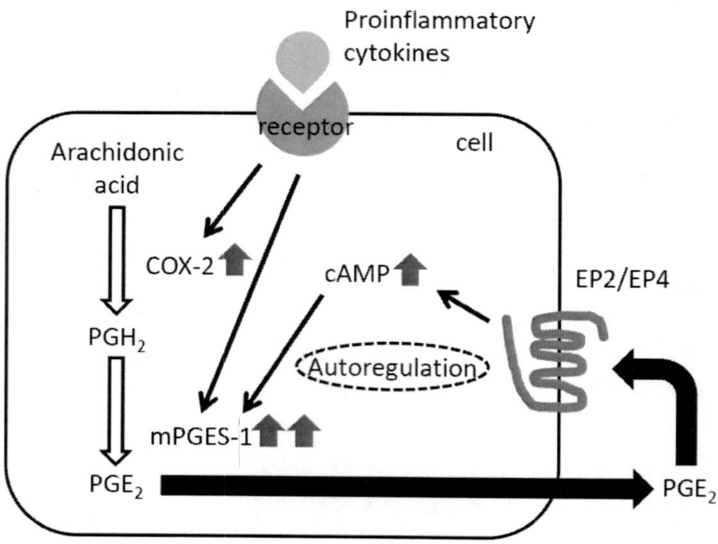

Figure 3. Pathway for positive regulation of mPGES-1 expression by PGE_2. COX-2 and mPGES-1 show coordinated induction in cultured synovial fibroblasts by stimulation with proinflammatory cytokines. At sites of inflammation, PGE_2 further enhances mPGES-1 expression associated with an increase of cyclic AMP (cAMP) *via* the EP2 and EP4 receptors.

Adiponectinand PGE_2 in RA

We found that the adipokine adiponectin increased COX-2 and mPGES-1 mRNA expression in synovial fibroblasts obtained from RA patients [53]. In contrast, leptin and resistin, even at levels up to 10-fold higher (100 ng/mL) than their serum concentrations, did not increase COX-2 and mPGES-1 mRNA expression by RA synovial fibroblasts (Figure 4).

To determine whether adiponectin increased the expression of PGESs and promoted to PGE_2 synthesis, we performed Western blotting analysis [53]. We found that adiponectin increased the expression of mPGES-1 protein in a concentration-dependent manner, whereas cPGES protein expression was unchanged.

Adiponectin enhanced PGE_2 production by RA synovial fibroblastsat concentrations found in human serum, while leptin and resistin had no significant effect on PGE_2 production (Figure 5). Small interfering RNA (siRNA) for the adiponectin receptor (AdipoR1 and AdipoR2) gene decreased adiponectin-induced PGE_2 production. Adiponectin-induced PGE_2 production was also significantly decreased by inhibitor of AMP-activated protein kinase (AMPK), suggesting thatwas at least partly induced by signal transduction via AdipoR1.

Yamauchi et al., [54] demonstrated that the peroxisome proliferator–activated receptorα (PPARα) signaling pathway exists downstream of AdipoR2. We found that an antagonist of the PPARα pathway reducesadiponectin-induced PGE_2 production by RA synovial fibroblasts.

NF-κB is known to play a central role in the regulation of inflammatory reactions in various cells [55]. Regarding PGE_2 production by RA synovial fibroblasts, NF-κB has an important role in the transcriptional regulation of COX-2 [56]. We demonstrated that adiponectin promotes NF-κB translocation in RA synovial fibroblasts, suggesting that adiponectin induces COX-2 expression in RA synovial fibroblasts via activation of NF-κB. Since the mPGES-1 promoter does not contain an NF-κB–responsive element, expression of mPGES-1 might be induced indirectly after activation of NF-κB [57], unlike COX-2. An increase of PGE_2 production due to COX-2 activation by adiponectin could lead to autocrine enhancement of mPGES-1 expression [52].

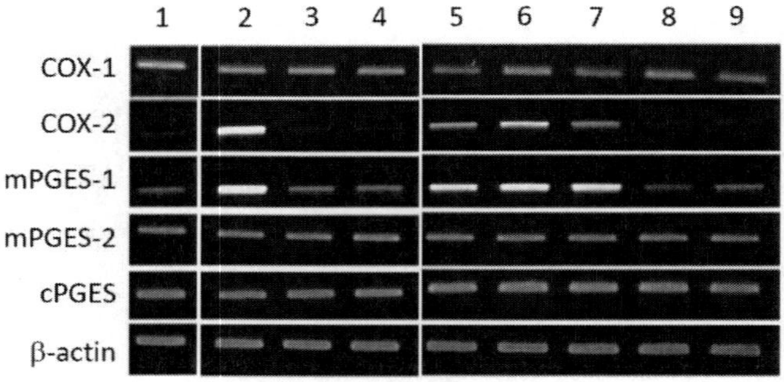

Figure 4. Effects of adipokines on COX and PGES mRNA expression. Synovial fibroblasts were incubated with IL-1β, adiponectin, leptin, or resistinat the indicated concentration for 18 h. Total RNA was isolated from cells and was subjected to the reverse transcription-polymerase chain reaction for each target enzyme and for β-actin. Lane 1, untreated; Lane 2, IL-1β (1 ng/mL); Lane 3, leptin (10 ng/mL); Lane 4, leptin (100 ng/mL); Lane 5, adiponectin (0.1 μg/mL); Lane 6, adiponectin (1 μg/mL); Lane 7, adiponectin (10 μg/mL); Lane 8, resistin (10 ng/mL); Lane 9, resistin (100 ng/mL).

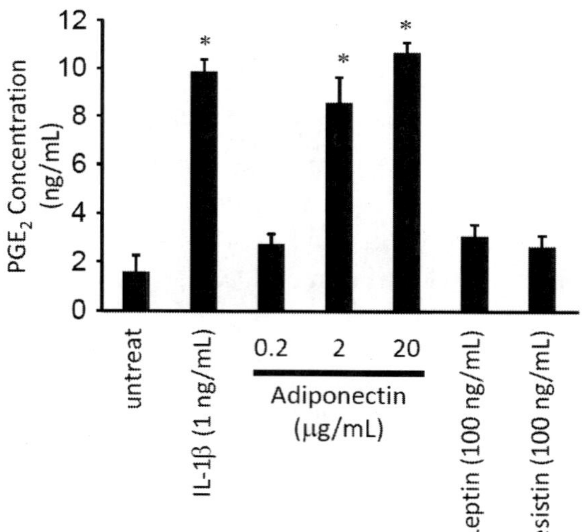

Figure 5. Prostaglandin E_2 production by RA synovial fibroblasts treated with adipokines. Synovial fibroblasts were incubated with IL-1β, adiponectin, leptin, or resistinat the indicated concentration for 18 h. After washing the cells, 3 μM arachidonic acidwas added for 30 min and then the PGE_2 concentration in the medium was measured by enzyme-linked immunosorbent assay. Bars show the mean and SD (n = 3). $*=p< 0.01$ versus untreat by Tukey's multiple comparison test.

Potential Role of AA metabolites in RA

Table1 shows various effects of PGE_2 that could be involved in joint inflammation and destruction. It can be suggested that reducing the bioactivity of PGE_2 is important to slow disease progression in RA.

We summarized the contribution of the PGE_2 biosynthesis pathway to arthritis in mice [58]. It was reported that mPGES-1-deficient mice showed marked reduction of inflammatory and histopahological changes such as pannus formation and joint erosion when used to create models of collagen-induced arthritis and collagen antibody-induced arthritis, which are animal models of inflammatory arthritis that resemble human RA [38, 39]. cPLA2-deficient mice also displayed considerable amelioration of both clinical and histological changes in these models [59-61].

These findings suggest that regulation of PGE_2 biosynthesis by both cytokines and adipokines may be a possible therapeutic target for RA.

REFERENCES

[1] Brennan, F.M. & McInnes, I.B. (2008). Evidence that cytokines play a role in rheumatoid arthritis. *J. Clin. Invest.*, 118, 3537-3545.

[2] Ahima, R.S. & Flier, J.S. (2000). Adipose tissue as an endocrine organ. *Trends Endocrinol. Metab.*, 11, 327-332.

[3] Meier, U. & Gressner, A.M. (2004). Endocrine regulation of energy metabolism: review of pathobiochemical and clinical chemical aspects of leptin, ghrelin, adiponectin, and resistin. *Clin. Chem.*, 50, 1511-1525.

[4] Tilg, H. & Moschen, A.R. (2006). Adipocytokines: mediators linking adipose tissue, inflammation and immunity. *Nat. Rev. Immunol.*, 6, 772-783.

[5] Oh, D.K;Ciaraldi, T; Henry, R.R. (2007). Adiponectin in health and disease. *Diabetes Obes. Metab.,* 9, 282-289.

[6] Yoshino, T; Kusunoki, N; Tanaka, N; Kaneko, K; Kusunoki, Y; Endo, H; Hasunuma, T; Kawai, S. (2011). Elevated serum levels of resistin, leptin, and adiponectin are associated with C-reactive protein and also other clinical conditions in rheumatoid arthritis. *Intern. Med.*, 50,269-275.

[7] Migita, K; Maeda, Y; Miyashita, T; Kimura, H; Nakamura, M; Ishibashi, H; Eguchi, K. (2006). The serum levels of resistin in rheumatoid arthritis patients. *Clin. Exp. Rheumatol.*, 24, 698-701.

[8] Senolt, L; Housa, D; Vernerová, Z; Jirásek, T; Svobodová, R;Veigl, D; Anderlová, K; Müller-Ladner, U; Pavelka, K; Haluzík, M. (2007). Resistin in rheumatoid arthritis synovial tissue, synovial fluid and serum. *Ann. Rheum. Dis.*,66, 458-463.

[9] Rho, Y. H; Solus, J; Sokka, T; Oeser, A; Chung, C. P; Gebretsadik, T; Shintani, A; Pincus, T; Stein, C.M. (2009). Adipocytokines are associated with radiographic joint damage in rheumatoid arthritis. *Arthritis Rheum*, 60, 1906-1914.

[10] Otero, M; Lago, R; Gomez, R; Lago, F; Dieguez, C; Gómez-Reino, J. J; Gualillo, O. (2006). Changes in plasma levels of fat-derived hormones adiponectin, leptin, resistin and visfatin in patients with rheumatoid arthritis., *Ann. Rheum. Dis.*, 65, 1198-1201.

[11] Targońska-Stepniak, B; Majdan, M; Dryglewska, M. (2008). Leptin serum levels in rheumatoid arthritis patients: relation to disease duration and activity. *Rheumatol. Int.,* 28, 585-591.

[12] Gunaydin, R; Kaya, T; Atay, A; Olmez, N; Hur, A; Koseoglu, M. (2006). Serum leptin levels in rheumatoid arthritis and relationship with disease activity. *South Med. J.*, 99, 1078-1083.

[13] Härle, P; Sarzi-Puttini, P; Cutolo, M; Straub, R.H. (2006). No change of serum levels of leptin and adiponectin during anti-tumour necrosis factor antibody treatment with adalimumab in patients with rheumatoid arthritis. *Ann. Rheum. Dis.*, 65, 970-971.

[14] Senolt, L; Pavelka, K; Housa, D; Haluzík, (2006). M. Increased adiponectin is negatively linked to the local inflammatory process in patients with rheumatoid arthritis. *Cytokine,* 35, 247-252.

[15] Popa, C; Netea, M. G; de Graaf, J; van den Hoogen, F. H; Radstake, T. R; Toenhake-Dijkstra, H; van Riel, P.L; van der Meer, J. W; Stalenhoef, A.F; Barrera, P. (2009). Circulating leptin and adiponectin concentrations during tumor necrosis factor blockade in patients with active rheumatoid arthritis. *J. Rheumatol.,* 36, 724-730.

[16] Harizi, H; Corcuff, J. B; Gualde, N. (2008). Arachidonic-acid-derived eicosanoids: roles in biology and immunopathology. *Trends in Molecular Medicine,* 14, 461-469.

[17] Kita, Y; Ohto, T; Uozumi, N; Shimizu, T. (2006). Biochemical properties and pathophysiological roles of cytosolic phospholipase A2s. *Biochim. Biophys. Acta.,* 1761, 1317-1322.

[18] Ghosh, M; Tucker, D.E; Burchett, S.A; Leslie, C.C. (2006). Properties of the Group IV phospholipase A2 family. *Prog. Lipid. Res.*, 45, 487-510.

[19] Tai, N; Kuwabara, K; Kobayashi, M; Yamada, K; Ono, T; Seno, K; Gahara, Y; Ishizaki, J; Hori, Y. (2010). Cytosolic phospholipase A2 alpha inhibitor, pyrroxyphene, displays anti-arthritic and anti-bone destructive action in a murine arthritis model. *Inflamm. Res.*, 59, 53-62.

[20] Xie, W. L; Chipman, J. G; Robertson, D. L; Erikson, R.L; Simmons, D.L. (1991). Expression of a mitogen-responsive gene encoding prostaglandin synthase is regulated by mRNA splicing. *Proc. Natl. Acad. Sci. U S A,* 88, 2692-2696.

[21] Kujubu, D.A; Fletcher, B. S; Varnum, B.C; Lim, R. W; Herschman, H.R. (1991). TIS10, a phorbol ester tumor promoter-inducible mRNA from Swiss 3T3 cells, encodes a novel prostaglandin synthase/ cyclooxygenase homologue. *J. Biol. Chem.*, 266, 12866-12872.

[22] Vane, J. R. (1996). Botting RM. Mechanism of action of anti-inflammatory drugs. *Scand. J. Rheumatol.*, 25, 9–21.

[23] Vane, J. R. (1971). Inhibition of prostaglandin synthesis as a mechanism of action for aspirin-like drugs. *Nature New Biol.*, 231, 232-235.

[24] Siegle, I; Klein, T; Backman, J. T; Saal, J. G; Nüsing, R.M; Fritz, P. (1998). Expression of cyclooxygenase 1 and cyclooxygenase 2 in human synovial tissue: differential elevation of cyclooxygenase 2 in inflammatory joint diseases. *Arthritis Rheum.*, 41, 122-129.

[25] Crofford, L.J; Wilder, R. L; Ristimäki, A.P; Sano, H; Remmers, E.F; Epps, H. R; Hla, T. (1994), Cyclooxygenase-1 and -2 expression in rheumatoid synovial tissues. Effects of interleukin-1 beta, phorbol ester, and corticosteroids. *J. Clin. Invest.*, 93, 1095-1101.

[26] Kalinski, P. (2012). Regulation of immune responses by prostaglandin E2. *J. Immunol.*, 188, 21-28.

[27] Egg, D; Günther, R; Herold, M; Kerschbaumer, F. (1980). Prostaglandins E2 and F2 alpha concentrations in the synovial fluid in rheumatoid and traumatic knee joint diseases. *Z Rheumatol.*, 39, 170-175.

[28] Kawai, S; Nishida, S; Kato, M; Furumaya, Y; Okamoto, R; Koshino, T; Mizushima, Y. (1998). Comparison of cyclooxygenase-1 and -2 inhibitory activities of various nonsteroidal anti-inflammatory drugs using human platelets and synovial cells. *Eur. J. Pharmacol.*, 347, 87-94.

[29] Tanioka, T; Nakatani, Y; Semmyo, N; Murakami, M; Kudo, I. (2000). Molecular identification of cytosolic prostaglandin E2 synthase that is functionally coupled with cyclooxygenase-1 in immediate prostaglandin E2 biosynthesis. *J. Biol. Chem.*, 275, 32775-32782.

[30] Jakobsson, P.J; Thoren, S; Morgenstern, R. (1999). Samuelsson B. Identification of human prostaglandin E synthase: a microsomal, glutathione-dependent, inducible enzyme, constituting a potential novel drug target. *Proc. Natl. Acad. Sci. USA,* 96, 7220-7225.

[31] Murakami, M; Naraba, H; Tanioka, T; Semmyo, N; Nakatani, Y; Kojima, F; Ikeda, T; Fueki, M; Ueno, A; Oh, S; Kudo, I. (2000). Regulation of prostaglandin E2 biosynthesis by inducible membrane-associated prostaglandin E2 synthase that acts in concert with cyclooxygenase-2. *J. Biol. Chem.*, 275, 32783-32792.

[32] Lazarus, M; Kubata, B. K; Eguchi, N; Fujitani, Y; Urade, Y; Hayaishi, O. (2002). Biochemical characterization of mouse micro-somalprostaglandin E synthase-1 and its colocalization with cyclooxygenase-2 in peritoneal macrophages. *Arch. Biochem. Biophys.*, 397, 336-341.

[33] Kojima, F; Naraba, H; Sasaki, Y; Okamoto, R; Koshino, T; Kawai, S. (2002). Coexpression of microsomal prostaglandin E synthase with cyclooxygenase-2 in human rheumatoid synovial cells. *J. Rheumatol.*, 29, 1836-1842.

[34] Stichtenoth, D. O; Thoren, S; Bian, H; Peters-Golden, M; Jakobsson, P. J; Crofford, L.J. (2001). Microsomal prostaglandin E synthase is regulated by proinflammatory cytokines and glucocorticoids in primary rheumatoid synovial cells. *J. Immunol.,* 167, 469-474.

[35] Han, R; Tsui, S; Smith, T.J. (2002). Up-regulation of prostaglandin E2 synthesis by interleukin-1beta in human orbital fibroblasts involves coordinate induction of prostaglandin-endoperoxide H synthase-2 and glutathione-dependent prostaglandin E2 synthase expression. *J. Biol. Chem.,* 277, 16355-16364.

[36] Saegusa, M; Murakami, M; Nakatani, Y; Yamakawa, K; Katagiri, M; Matsuda, K; Nakamura, K; Kudo, I; Kawaguchi, H. (2003). Contribution of membrane-associated prostaglandin E2 synthase to bone resorption. *J. Cell Physiol.,* 197, 348-356.

[37] Kojima, F; Naraba, H; Miyamoto, S; Beppu, M; Aoki, H; Kawai, S. (2004). Membrane-associated prostaglandin E synthase-1 is upregulated by proinflammatory cytokines in chondrocytes from patients with osteoarthritis. *Arthritis Res. Ther.*, 6, R355-R365.

[38] Trebino, C. E; Stock, J. L; Gibbons, C. P; Naiman, B. M; Wachtmann, T. S; Umland, J. P; Pandher, K; Lapointe, J. M; Saha, S; Roach, M.L; Carter, D; Thomas, N. A; Durtschi, B. A; McNeish, J. D; Hambor, J. E; Jakobsson, P.J; Carty, T. J; Perez, J. R; Audoly, L. P. (2003). Impaired

inflammatory and pain responses in mice lacking an inducible prostaglandin E synthase. *Proc. Natl. Acad. Sci. USA*, 100, 9044-9049.

[39] Kamei, D; Yamakawa, K; Takegoshi, Y; Mikami-Nakanishi, M; Nakatani, Y; Oh-Ishi, S; Yasui, H; Azuma, Y; Hirasawa, N; Ohuchi, K; Kawaguchi, H; Ishikawa, Y; Ishii, T; Uematsu, S; Akira, S; Murakami, M; Kudo, I. (2004). Reduced pain hypersensitivity and inflammation in mice lacking microsomal prostaglandin E synthase-1. *J. Biol. Chem.*, 279, 33684-33695.

[40] Mabuchi, T; Kojima, H; Abe, T; Takagi, K; Sakurai, M; Ohmiya, Y; Uematsu, S; Akira, S; Watanabe, K; Ito, S. (2004). Membrane-associated prostaglandin E synthase-1 is required for neuropathic pain. *Neuroreport,* 15, 1395-1398.

[41] Yamagata, K; Matsumura, K; Inoue, W; Shiraki, T; Suzuki, K; Yasuda, S; Sugiura, H; Cao, C; Watanabe, Y; Kobayashi, S. (2001). Coexpression of microsomal-type prostaglandin E synthase with cyclooxygenase-2 in brain endothelial cells of rats during endotoxin-induced fever. *J. Neurosci.,* 21, 2669-2677.

[42] Ivanov, A. I; Pero, R.S; Scheck, A. C; Romanovsky, A.A. (2002). Prostaglandin E(2)-synthesizing enzymes in fever: differential transcriptional regulation. *Am. J. Physiol. Regul. Integr. Comp. Physiol.*, 283, R1104-R1117.

[43] Engblom, D; Saha, S; Engström, L; Westman, M; Audoly, L. P; Jakobsson, P.J; Blomqvist, A. (2003). Microsomal prostaglandin E synthase-1 is the central switch during immune- induced pyresis. *Nat. Neurosci.*, 6, 1137-1138.

[44] Ivanov, A. I.& Romanovsky, A.A. (2004). Prostaglandin E2 as a mediator of fever: synthesis and catabolism. *Front Biosci.*, 9, 1977-1993.

[45] Filion, F; Bouchard, N; Goff, A. K; Lussier, J. G; Sirois, J. (2001). Molecular cloning and induction of bovine prostaglandin E synthase by gonadotropins in ovarian follicles prior to ovulation in vivo. *J. Biol. Chem.*, 276, 34323-34330.

[46] Lazarus, M; Munday, C. J; Eguchi, N; Matsumoto, S; Killian, G.J; Kubata, B.K; Urade, Y. (2002). Immunohistochemical localization of microsomal PGE synthase-1 and cyclooxygenases in male mouse reproductive organs. *Endocrinology*, 143, 2410-2419.

[47] Satoh, K; Nagano, Y; Shimomura, C; Suzuki, N; Saeki, Y; Yokota, H. (2000). Expression of prostaglandin E synthase mRNA is induced in beta-amyloid treated rat astrocytes. *Neurosci. Lett.,* 283, 221-223.

[48] Cipollone F., Prontera C., Pini B. et al., Over expression of functionally coupled cyclooxygenase-2 and prostaglandin E synthase in symptomatic atherosclerotic plaques as a basis of prostaglandin E(2)-dependent plaque instability. *Circulation* (2001) 104 921–927.

[49] Cipollone, F; Prontera, C; Pini, B; Marini, M; Fazia, M; De Cesare, D; Iezzi, A; Ucchino, S; Boccoli, G; Saba, V; Chiarelli, F; Cuccurullo, F; Mezzetti, A. (2003). The receptor RAGE as a progression factor amplifying arachidonate-dependent inflammatory and proteolytic response in human atherosclerotic plaques: role of glycemic control. *Circulation,* 108, 1070-1077.

[50] Kamei, D; Murakami, M; Nakatani, Y; Ishikawa, Y; Ishii, T; Kudo, I. (2003). Potential role of microsomal prostaglandin E synthase-1 in tumorigenesis. *J. Biol. Chem.,* 278, 19396-19405.

[51] Murakami, M; Nakashima, K; Kamei, D; Masuda, S; Ishikawa, Y; Ishii, T; Ohmiya, Y; Watanabe, K; Kudo, I. (2003). Cellular prostaglandin E2 production by membrane-bound prostaglandin E synthase-2 via both cyclooxygenases-1 and -2. *J. Biol. Chem.*, 278, 37937-37947.

[52] Kojima, F; Naraba, H; Sasaki, Y; Beppu, M; Aoki, H; Kawai, S. (2003). Prostaglandin E2 is an enhancer of interleukin-1beta-induced expression of membrane-associated prostaglandin E synthase in rheumatoid synovial fibroblasts. *Arthritis Rheum.*, 48, 2819-2828.

[53] Kusunoki, N; Kitahara, K; Kojima, F; Tanaka, N; Kaneko, K; Endo, H; Suguro, T; Kawai, S. (2010). Adiponectin stimulates prostaglandin E(2) production in rheumatoid arthritis synovial fibroblasts. *Arthritis Rheum.,* 62, 1641-1649.

[54] Yamauchi, T; Nio, Y; Maki, T; Kobayashi, M; Takazawa, TIwabu, M; Okada-Iwabu, M; Kawamoto, S; Kubota, N; Kubota, T; Ito, Y;Kamon, J; Tsuchida, A; Kumagai, K; Kozono, H; Hada, Y; Ogata, H; Tokuyama, K; Tsunoda, M; Ide, T; Murakami, K; Awazawa, M; Takamoto, I; Froguel, P; Hara, K;Tobe, K; Nagai, R; Ueki, K; Kadowaki, T. (2007). Targeted disruption of AdipoR1 and AdipoR2 causes abrogation of adiponectin binding and metabolic actions. *Nat. Med.,* 13, 332-339.

[55] Barnes, P.J. & Karin, M. (1997). Nuclear factor-kappa B: a pivotal transcription factor in chronic inflammatory diseases. *N. Engl. J. Med.*, 336, 1066-1071.

[56] Crofford, L.J; Tan, B; McCarthy, C. J; Hla, T. (1997). Involvement of nuclear factor kappa B in the regulation of cyclooxygenase-2 expression by interleukin-1 in rheumatoid synoviocytes. *Arthritis Rheum.*, 40, 226-236.

[57] Fahmi, H. (2004). mPGES-1 as a novel target for arthritis. *Curr. Opin. Rheumatol.*,16, 623-627.

[58] Kojima, F; Kato, S; Kawai, S. (2005). Prostaglandin E synthase in the pathophysiology of arthritis. *Fundam. Clin. Pharmacol.*, 19, 255-261.

[59] Hegen, M; Sun, L; Uozumi, N; Kume, K; Goad, M.E; Nickerson-Nutter, C.L; Shimizu, T; Clark, J.D. (2003). Cytosolic phospholipase A2alpha-deficient mice are resistant to collagen-induced arthritis. *J. Exp. Med.*, 197, 1297-302.

[60] Myers, L.K; Kang, A. H; Postlethwaite, A. E; Rosloniec, E. F; Morham, S. G; Shlopov, B. V; Goorha, S; Ballou, L.R. (2000). The genetic ablation of cyclooxygenase 2 prevents the development of autoimmune arthritis. *Arthritis Rheum.*,43, 2687-2693.

[61] McCoy, J.M; Wicks, J. R; Audoly, L.P. (2002). The role of prostaglandin E2 receptors in the pathogenesis of rheumatoid arthritis. *J. Clin. Invest.*, 110, 651-658.

[62] Moriyama T, Higashi T, Togashi K, Iida T, Segi E, Sugimoto Y, Tominaga T, Narumiya S, Tominaga M. Sensitization of TRPV1 by EP1 and IP reveals peripheral nociceptive mechanism of prostaglandins. *Mol. Pain.* 2005 Jan 17;1:3.

[63] Villarreal CF, Sachs D, Cunha FQ, Parada CA, Ferreira SH. The role of Na(V)1.8 sodium channel in the maintenance of chronic inflammatory hypernociception. *Neurosci. Lett.* 2005 Sep 30;386(2):72-7.

[64] Amaya F, Wang H, Costigan M, Allchorne AJ, Hatcher JP, Egerton J, Stean T, Morisset V, Grose D, Gunthorpe MJ, Chessell IP, Tate S, Green PJ, Woolf CJ. The voltage-gated sodium channel Na(v)1.9 is an effector of peripheral inflammatory pain hypersensitivity. *J. Neurosci.* 2006 Dec 13;26(50):12852-60.

[65] Wang C, Gu Y, Li GW, Huang LY.A critical role of the cAMP sensor Epac in switching protein kinase signalling in prostaglandin E2-induced potentiation of P2X3 receptor currents in inflamed rats. *J. Physiol.* 2007 Oct 1;584(Pt 1):191-203.

[66] Nishihara I, Minami T, Watanabe Y, Ito S, Hayaishi O. Prostaglandin E2 stimulates glutamate release from synaptosomes of rat spinal cord. *Neurosci. Lett.* 1995 Aug 18;196(1-2):57-60.

[67] Nicol GD, Klingberg DK, Vasko MR. Prostaglandin E2 increases calcium conductance and stimulates release of substance P in avian sensory neurons. *J. Neurosci.* 1992 May;12(5):1917-27.

[68] Kohno T, Wang H, Amaya F, Brenner GJ, Cheng JK, Ji RR, Woolf CJ. Bradykinin enhances AMPA and NMDA receptor activity in spinal cord

dorsal horn neurons by activating multiple kinases to produce pain hypersensitivity. *J. Neurosci.* 2008 Apr 23;28(17):4533-40.

[69] Harvey RJ, Depner UB, Wässle H, Ahmadi S, Heindl C, Reinold H, Smart TG, Harvey K, Schütz B, Abo-Salem OM, Zimmer A, Poisbeau P, Welzl H, Wolfer DP, Betz H, Zeilhofer HU, Müller U. GlyR alpha3: an essential target for spinal PGE2-mediated inflammatory pain sensitization. *Science.* 2004 May 7;304(5672):884-7.

[70] Morimoto K, Shirata N, Taketomi Y, Tsuchiya S, Segi-Nishida E, Inazumi T, Kabashima K, Tanaka S, Murakami M, Narumiya S, Sugimoto Y. Prostaglandin E2-EP3 Signaling Induces Inflammatory Swelling by Mast Cell Activation. *J. Immunol.* 2013 Dec 16. [Epub ahead of print]

[71] Foudi N, Gomez I, Benyahia C, Longrois D, Norel X. Prostaglandin E2 receptor subtypes in human blood and vascular cells. *Eur. J. Pharmacol.* 2012 Nov 15;695(1-3):1-6.

[72] Suzawa T, Miyaura C, Inada M, Maruyama T, Sugimoto Y, Ushikubi F, Ichikawa A, Narumiya S, Suda T. The role of prostaglandin E receptor subtypes (EP1, EP2, EP3, and EP4) in bone resorption: an analysis using specific agonists for the respective *EPs. Endocrinology.* 2000 Apr;141(4):1554-9.

[73] Li X, Okada Y, Pilbeam CC, Lorenzo JA, Kennedy CR, Breyer RM, Raisz LG. Knockout of the murine prostaglandin EP2 receptor impairs osteoclastogenesis in vitro. *Endocrinology.* 2000 Jun;141(6):2054-61.

[74] Li X, Pilbeam CC, Pan L, Breyer RM, Raisz LG. Effects of prostaglandin E2 on gene expression in primary osteoblastic cells from prostaglandin receptor knockout mice. *Bone.* 2002 Apr;30(4):567-73.

[75] Li X, Tomita M, Pilbeam CC, Breyer RM, Raisz LG. Prostaglandin receptor EP2 mediates PGE2 stimulated hypercalcemia in mice in vivo. Prostaglandins *Other Lipid Mediat.* 2002 Apr;67(3-4):173-80.

[76] Moreno-Rubio J, Herrero-Beaumont G, Tardio L, Alvarez-Soria MA, Largo R. Non steroidalanti inflammatory drugs and prostaglandin E(2) modulate the synthesis of osteoprotegerin and RANKL in the cartilage of patients with severe knee osteoarthritis. *Arthritis Rheum.* 2010 Feb;62(2):478-88.

[77] Brändström H, Jonsson KB, Ohlsson C, Vidal O, Ljunghall S, Ljunggren O. Regulation of osteoprotegerin mRNA levels by prostaglandin E2 in human bone marrow stroma cells. *Biochem. Biophys. Res. Commun.* 1998 Jun 18;247(2):338-41.

[78] Suda K, Udagawa N, Sato N, Takami M, Itoh K, Woo JT, Takahashi N, Nagai K. Suppression of osteoprotegerin expression by prostaglandin E2 is crucially involved in lipopolysaccharide-induced osteoclast formation. *J. Immunol.* 2004 Feb 15;172(4):2504-10.

[79] Notoya K, Jovanovic DV, Reboul P, Martel-Pelletier J, Mineau F, Pelletier JP. The induction of cell death in human osteoarthritis chondrocytes by nitric oxide is related to the production of prostaglandin E2 via the induction of cyclooxygenase-2. *J. Immunol.* 2000 Sep 15;165(6):3402-10.

[80] Miyaura C, Inada M, Suzawa T, Sugimoto Y, Ushikubi F, Ichikawa A, Narumiya S, Suda T. Impaired bone resorption to prostaglandin E2 in prostaglandin E receptor EP4-knockout mice. *J. Biol. Chem.* 2000 Jun 30;275(26):19819-23.

[81] Kim CH, Park YG, Noh SH, Kim YK.PGE2 induces the gene expression of bone matrix metalloproteinase-1 in mouse osteoblasts by cAMP-PKA signaling pathway. *Int. J. Biochem. Cell Biol.* 2005 Feb;37(2):375-85.

[82] Attur M, Al-Mussawir HE, Patel J, Kitay A, Dave M, Palmer G, Pillinger MH, Abramson SB. Prostaglandin E2 exerts catabolic effects in osteoarthritis cartilage: evidence for signaling via the EP4 receptor. *J. Immunol.* 2008 Oct 1;181(7):5082-8.

[83] Miles RR, Sluka JP, Halladay DL, Santerre RF, Hale LV, Bloem L, Thirunavukkarasu K, Galvin RJ, Hock JM, Onyia JE.ADAMTS-1: A cellular disintegrin and metalloprotease with thrombospondin motifs is a target for parathyroid hormone in bone. *Endocrinology.* 2000 Dec;141(12):4533-42.

INDEX

#

A

B

C

D

E

F

I

Q

R

S

T

U

V

W

X

Y